Advanced Information Processing Technology
A series edited by Tadao Saito

Volume 1

Domain Oriented Systems Development - principles and approaches
Edited by K. Itoh, T. Hirota, S. Kumagai and H. Yoshida (1998)

Volume 2

Designing Communications and Collaboration Support Systems
Edited by Y. Matsushita (1999)

Volume 3

Information Networking in Asia
Edited by H. Higaki, Y. Shibata and M. Takizawa (2001)

Volume 4

Advanced Lisp Technology
Edited by T. Yuasa and H. G. Okuno (2002)

Volume 5

Nontraditional Database Systems
Edited by Y. Kambayashi, M. Kitsuregawa, A. Makinouchi, S. Uemura, K. Tanaka and Y. Masunaga (2002)

Nontraditional Database Systems

Edited by

Yahiko Kambayashi
Kyoto University

Masaru Kitsuregawa
The University of Tokyo

Akifumi Makinouchi
Kyushu University

Shunsuke Uemura
Nara Institute of Science and Technology

Katsumi Tanaka
Kobe University

Yoshifumi Masunaga
Ochanomizu University

The Information Processing Society of Japan

CRC Press
Taylor & Francis Group
Boca Raton London New York

CRC Press is an imprint of the
Taylor & Francis Group, an **informa** business
A TAYLOR & FRANCIS BOOK

CRC Press
Taylor & Francis Group
6000 Broken Sound Parkway NW, Suite 300
Boca Raton, FL 3487-2742

ISBN-13: 978-0-415-30206-7 (hbk)

Visit the Taylor & Francis Web site at
http://www.taylorandfrancis.com

and the CRC Press Web site at
http://www.crcpress.com

Publisher's note

This book has been prepared from camera-ready-copy provided by the editors

Every effort has been made to ensure that the advice and information in this book is true and accurate at the time of going to press. However, neither the publisher nor the authors can accept any legal responsibility or liability for any errors or omissions that may be made. In the case of drug administration, any medical procedure or the use of technical equipment mentioned within this book, you are strongly advised to consult the manufacturer's guidelines.

British Library Cataloguing in Publication Data
A catalogue record for this book is available from the British Library

Library of Congress Cataloging in Publication Data
A catalog record for this book has been requested

Contents

Series Foreword

The Information Processing Society of Japan (IPSJ) is the top academic institution in the information processing field in Japan. It has about thirty thousand members and promotes a variety of research and development activities covering all aspects of information processing and computer science.

One of the major activities of the society is publication of its transactions containing papers covering all the fields of information processing, including fundamentals, software, hardware, and applications. Some of the papers are published in English, but because the majority are in Japanese, the transactions are not suitable for non-Japanese wishing to access advanced information technology in Japan. IPSJ therefore decided to publish a book series titled "Advanced Information Technology in Japan."

The series consists of independent books, each including a collection of top quality papers, from mainly Japanese sources of a selected area in information technology. The book titles were chosen by the International Publication Committee of IPSJ so that they enable easy access to information technology form international readers. Each book contains original papers and/or those updated or translated from original papers appearing in the transactions or internationally qualified meetings. Survey papers to aid understanding of the technology in Japan in a particular area are also included.

As the chairman of the International Publication Committee of IPSJ, I sincerely hope that the books in the series will improve communication between Japanese and non-Japanese specialists for their mutual benefit.

Tadao Saito

Series Editor

Chairman
International Publication Committee
the Information Processing Society of Japan

Preface

This book contains selected research results on non-traditional database systems from Japanese database research community. In traditional database research, the main emphases is how to process a large amount of fixed structured data with minimum amount of selected resources (processing time, communication overhead, memory size, storage size, etc.).

We will focus on non-traditional types of (1) applications, (2) data types, (3) systems, (4) environment, together with (5) high-performance architecture to support non-traditional applications.

For non-traditional applications, we have selected topics on real-world models, interaction between real world and virtual world, and support of groupwork among users. For the first two applications we have to handle large among of three dimensional data. Database support for groupwork is essential, since sharing of information of the current and the past is required.

For non-traditional data types, continuous media (especially video) is considered together with human impression (Kansei) representation. For non-traditional systems, we selected systems for the above applications and data types. More specifically, systems to support real world data, video data and GIS applications are shown. For non-traditional environment, web is selected and handling of heterogeneous collection of documents including XML presentation are discussed.

For all these new problems we need high-performation processing. Three examples, web mining, data engineering and object processing are discussed.

As non-traditional databases are getting important, we really hope this volume will contributed to enlarge research communities engaging such influential topics.

Prior to the work on this book we had a three-year project on advanced database research started in April 1996. Major results of the projects are reported in the Advanced Database Research and Development Series published by World Scientific.

Vol.7 : Cooperative database and Applications ('96) *(ISBN 981-02-3161-X)*

Vol.8 : Digital Media Information Base ('97) *(ISBN 981-02-3437-6)*

Vol.9 : Advanced Database Systems for Integration of Media and User Environment ('98) *(ISBN981-02-3436-8)*

One year after the project, a symposium was held in Kyoto, which includes extended research results from the project together with research papers with related topics prepared by foreign researchers. The proceedings was published by IEEE Computer Society Press.

Database Application in Non-Traditional Environment '99 *(ISBN 0-7695-0496-5)*

Readers who are interested in the topics discussed here may find related preliminary results in these volumes.

In real database applications still traditional ones are dominant, we hope the topics covered in this book are getting important basis for database applications of the 21st century.

Yahiko Kambayashi
Masaru Kitsuregawa
Akifumi Makinouchi
Shunsuke Uemura
Katsumi Tanaka
Yoshifumi Masunaga

The Authors

Yahiko Kambayashi, received the B.E., M.E., and Ph. D. degrees from Kyoto University, Kyoto, Japan, in 1965, 1967, and 1970, respectively. In 1984 he became a Professor of Kyushu University, and since 1990 he has been a Professor of Kyoto University. He was a member of VLDB endowment, a vice-chair of ACM Tokyo/Japan Chapter, and members of editorial board of several international journals. He is a fellow of IEEE. IPSJ and IEICE, and a member of ACM.

Masaru Kitsuregawa, received the B.E. degree in electronics engineering in 1978, and the Dr.Eng. degree in information engineering in 1983 from The University of Tokyo. In 1983 joined Institute of Industrial Science, the University of Tokyo as a lecturer. He is currently a full professor at the University of Tokyo and a director of Conceptual Information Processing Research Center. His research interests include parallel computer architecture and database engineering. He has published around 150 referenced journal/conference papers. He has served a member of program committee for a lot of major international conferences such as ACM SIGMOD, IEEE ICDE, VLDB etc. and served a program co-chair of IEEE International Conference on Data Enginnering in 1999 and a co-general chair of Pacific Asia Knowledge Discovery and Data Mining Conference in 2000. He is serving(served) an editor of VLDB journal, DAPD(Distributed and Parallel Database Journal), New Generation Computing, IEICE journal etc. He is a trustee member of VLDB endowment and a member of IEEE Technical Committee of Data Engineering. He was the chairman of the technical group on data engineering in IEICE Japan. He is currently the chair of ACM SIGMOD Japan chapter.

Akifumi Makinouchi received his B.E. degree from Kyoto University, Japan, in 1967, Docteur-Ingereur degree from Univercite de Grenoble, France, in 1979, and D.E. degree from Kyoto University, Japan, in 1988. Since 1996, he has been with the Graduate School of Information Science and Electrical Engineering, Kyushu University, Japan, where he is a professor. He is a member of IPSJ, ACM, and IEEE.

Shunsuke Uemura, received his B.E. M.E. and Dr.Eng. degrees from Kyoto University in 1964, 1966 and 1975 respectively. In 1966, he joined the research staff of the Electrotechnical Laboratory (ETL), MITI, Japan. Meantime from 1970 to 1971 he was a Visiting Researcher at the Electronic Systems Laboratory, Massachusetts Institute of Technology. From 1988 to 1993, he was a professor at Department of Engineering, Tokyo University of Agriculture and Technology. Since 1993, he has been a professor at the Graduate School of Information Science, Nara Institute of Science and Technology. His current research interests include database systems and natural language processing. He is a fellow of Information Processing Society of Japan. He is a member of the ACM and IEEE Computer Society. He is currently serving as the area editor for multimedia and databases of the IEEE Computer magazine.

Katsumi Tanaka He received his B.E., M.E., and D.Eng. degrees in information science from Kyoto University, in 1974, 1976, and 1981, respectively. Since 1994, he is a professor of the Department of Computer and Systems Engineering and since 1997, he is a professor of the Division of Information and Media Sciences, Graduate School of Science and Technology, Kobe University. His research interests include object-oriented, multimedia and historical databases and multimedia information systems. He is a member of the ACM, IEEE Computer Society and the Information Processing Society of Japan.

Yoshifumi Masunaga, is a professor of the Department of Information Science of the Faculty of Science at Ochanomizu University, Tokyo, Japan. He graduated and received the Doctor of Engineering degree in Electrical Communication from Tohoku University, Japan in 1970. He was a research member of the Computer Science Department of the International Institute of Applied Systems Analysis (IIASA), Laxenburg, Austria from 1975 to 1977. He was a visiting scientist of the Computer Science Department of IBM San Jose Research Laboratory, San Jose, USA from 1982 to 1983, where he belonged to System R* Project. He was a professor of University of Library and Information Science, Japan from 1983 to 1999. He has been served as Chair of the Special Interest Group on Database Systems of the Information Processing Society of Japan (IPSJ SIGDBS) and Chair of the Japan Chapter of the ACM Special Interest Group on Database Systems (ACM SIGMOD Japan). He was an Orbiter of the Board of the Directors of IPSJ. He is also a member of the Steering Committee of DASFAA, that is the international database conference initiated in 1989 to promote database research and development activities in Asian/Australasian countries. He was an Associate Editor of the ACM Transactions on Database Systems (ACM TODS). He is also an Editor of the IPSJ Transactions on Databases (IPSJ TOD). He has published extensively in technical journals and conferences in Japanese and English. He has also authored texts on Relational Databases, which are widely used in Japanese universities. He is a member of ACM, IEEE Computer Society, and the Institute of Electronics, Information, and Communication Engineers of Japan (IEICE), and a fellow of IPSJ. His current research interests include multimedia database systems, virtual world database systems, moving object database systems, and data mining.

Ryota Abe received the B.E, and M.E. degrees from Tokyo Institute of Technology in 1999, and 2001, respectively. He is now a researcher at Internet Laboratories of Sony Corporation. His research interests include distributed data engineering and Internet security.

Toshiyuki Amagasa, received the B.E. M.E. and Ph.D degrees from Gunma University in 1994, 1996 and 1999, respectively. Since 1999, he as been an Research

Associate at the Graduate School of Information Science, Nara Institute of Science and Technology. His research interests include temporal and multimedia databases. He is a member of IEICE, IPSJ, ACM and IEEE CS.

Hirofumi Amano received a doctoral degree in Computer Science from Kyushu University in 1991 and joined the Department of Computer Science and Communication Engineering. Since 1994, he has been an associate professor at Computer Center, Kyushu University (the computer center was reorganized into Computing and Communications Center in 2000). His research interests include database programming languages and parallel processing for object-oriented databases. He is a member of IPSJ, IEICE, and ACM.

Masatoshi Arikawa is an associate professor of the Center of Spatial Information Science at the University of Tokyo. His research interests include databases, geographic information systems, virtual reality, user interface, object-oriented programming, and cartography. He received the B.E., M.E., and Ph.D. degrees in computer science and communication engineering from Kyushu University, Fukuoka, Japan, in 1986, 1988 and 1992.

Hiroshi Arisawa received the B.S. from University of Tokyo in 1972 and Ph.D. degree (Dr.Eng.) from Kyoto University in 1986 respectively. He was engaged in Fujitsu Ltd. from 1973 to 1974. In 1975, he joined to Yokohama National University. Now hi is a professor in the Faculty of Environment and Information Sciences. His research interest covers: theory of database design and multimedia database systems. He also prposed the "Real World Database" as the foundation of next generation database concept including spatio temporal objects. From 1986 to 1989, he was the chief of editorial WG of "Journal of Information Processing." He was also a visiting associate professor of Oregon State University from February to December in 1991.

Hiroshi Imai, Hiroshi Imai obtained B.Eng. in Mathematical Engineering, and M.Eng. and D.Eng. in Information Engineering, University of Tokyo in 1981, 1983 and 1986, respectively. In 1986~1990, He was an associate professor of Department of Computer Science and Communication Engineering, Kyushu University. He was also a visiting associate professor at School of Computer Science, McGill University in 1987 and a visiting scientist at IBM T. J. Watson Research Center in 1988. Since 1990, he has been an associate professor at Department of Information Science, University of Tokyo. His research interests include algorithms, computational geometry, and optimization. He is a member of IEICE, IPSJ, OR Soc. Japan and ACM.

Kunihiko Kaneko received a doctoral degree from Kyushu University in 1995, and joined the Department of Computer Science and Communication Since 1999, he has been an associate professor at Department of Intelligent Systems, Kyushu University. His research interests include parallel processing for object-oriented databases, transaction processing, and multimedia database. He is a member of IPSJ, IEICE, IEEE and ACM.

Hiroyuki Kitagawa received the B.Sc. degree in physics and the M.Sc. and D.Sc. degrees in computer science, in 1978, 1980, and 1987, respectively, all from the University of Tokyo. He is a full professor at Institute of Information Sciences and Electronics, University of Tsukuba, Japan. His research interests include data integration, semistructured data, structured documents, multimedia, and query processing.

Yasushi Kiyoki received his B.E., M.E. and Ph.D. degrees in electrical engineering from Keio University in 1978, 1980 and 1983, respectively. In 1983 and 1984, he worked at Electrical and Communication Labratory of NTT. From 1984 to 1996, he was with Institute of Information Sciences and Electronics, Univ. of Tsukuba, as an assistant professor and then an associate professor. Since 1996, he has been with Department of Environmental Information at Keio University, where he is currently a professor. His research addresses multidatabase systems, knowledge base systems, semantic associative processing, and multimedia database systems. He serves as the chair on Special Interest Group of Database Systems and as the editor-in-chief on Transactions on Databases in Information Processing Society of Japan. He also serves as the program chair for the 7th International Conference on Database Systems for Advanced Applications.

Takeo Kunishima received his B.E., M.E., and Dr. Eng. degrees in information science, in 1989, 1991, and 1997, respectively, all from Kyoto University. He joined the faculty at Nara Institute for Science and Technology and at Okayama Prefectural University in 1997. He is now Associate Professor in the Faculty of Computer Science and System Engineering.

Bojiang Liu received B.E. degree in 1982 from Xidian University, China, M.E. and Dr. Eng. degrees in engineering science in 1991 and 1995, respectively, from Osaka University. He joined the faculty at Osaka University in 1992 and at Okayama University of Science in 1996. He is now Associate Professor of the Faculty of Informatics.

Qiang Ma He received his B.E. from Hiroshima Prefectual University in 1998, and his M.E. in Information and Systems Engineering from Kobe University in 2000. Currently, he is a Ph.D. candidate student in Information and Media Sciences of Kobe University.

Akiyoshi Matono received B.E. degree in information engineering in 2000 from Okayama Prefectural University. He is now a student in the Graduate School of System Engineering.

Atsuyuki Morishima received the B.S., M.E. and D.E. degrees in computer science from University of Tsukuba, Japan, in 1993, 1995, and 1998, respectively. He is a research fellow of the Japan Society for the Promotion of Science, and currently working at Institute of Information Sciences and Electronics, University of Tsukuba. His current research interests include conversion and integration of heterogeneous data, Web management/applications, and user interfaces for data-intensive information systems.

Akiyo Nadamoto She received her B.E. in Electronics from The Science University of Tokyo in 1987. Since 1995, she has been affiliated with Kansai Research Institute,

Since 1998, she has been a Ph.D. candidate student in Information and Media Sciences of Kobe University. Her research interests include visualaization, multimedia systems, web information systems and databases.

Shojiro Nishio, received his B.E., M.E., and Dr.E. degrees from Kyoto University, Kyoto, Japan, in 1975, 1977, and 1980, respectively. From 1980 to 1988, he was with the Department of Applied Mathematics and Physics, Kyoto University. In October 1988, he joined the faculty of the Department of Information and Computer Sciences, Osaka University, Osaka, Japan. Since August 1992, he has been a full professor in the Department of Information Systems Engineering of Osaka University. He has been serving as the director of Cybermedia Center of Osaka University since April 2000. His current research interests include database systems, multimedia systems, and distributed computing systems. He has served on the editorial board of *IEEE Transactions on Knowledge and Data Engineering,* and is currently involved in the editorial boards of *ACM Transactions on Internet Technology, Data & Knowledge Engineering, New Generation Computing, International Journal of Information Technology, Data Mining and Knowledge Discovery,* and *The VLDB Journal.* He is a member of eight learned societies, including ACM and IEEE.

Takefumi Ogawa, received his B.E. and M.E. degrees in Information Systems Engineering from Osaka University, Osaka, Japan, in 1997 and 1999, respectively. Currently, he is a Research Associate of the Infomedia Education Division, Cybermedia Center, Osaka University. He is a member of IEEE, IEICE, IPSJ, and VRSJ. His research interests include virtual reality systems and augmented reality systems.

Ayumi Ohsugi, received her B.S. degree in Information Science from Ochanomizu University, Tokyo, Japan, in 2000. Currently, she is a master course student of the Graduate School of Humanities and Sciences, Ochanomizu University. Her research interests include spatio-temporal database systems, particularly the integration of a virtual reality system and an object-oriented database system.

Iko Pramudiono, received his B.E. in electronical engineering in 1998 and his M.E. degree in electro information engineering in 2000, all from The University of Tokyo. Currently, a doctor candidate there engaging in web mining research.

Kozue Satoh, received her B.S. degree in Information Science from Ochanomizu University, Tokyo, Japan, in 2000. Currently, she is a master course student of the Graduate School of Humanities and Sciences, Ochanomizu University. Her research interests include spatio-temporal database systems, particularly the integration of a virtual reality system and an object-oriented database system.

Kazutoshi Sumiya He received his B.E. and M.E. degrees in Instrumentation Engineering, and D.Eng degrees in Information and Media Sciences from Kobe University in 1986, 1988, and 1998, respectively. He was affiliated with Matsushita Electric Industrial Co. Ltd. during 1988-1999. Currently, he is a lecturer of Dept. of Computer and Systems Engineering and Research Center for Urban Safety and Security, Kobe University. His research interests include data broadcast, WWW, hyperText and hypermedia, multimedia databases, software development systems, and object-oriented databases. He is a member of IPSJ (SIGDBS), ITE, GISA ACM, and IEEE Computer Society.

Takayuki Tamura, received the B.E. degree in electronic engineering in 1991 and the M.E. and Dr.Eng. degrees in information engineering in 1993 and 1998, respectively, all from the University of Tokyo. Until 1998, he had been engaged in development of parallel relational database machines at Institute of Industrial Science, the University of Tokyo. He is currently a researcher at Mitsubishi Electric Corporation.

Takashi Tomii received the B.E., M.E. and Ph.D. degrees in electrical and computer engineering from Yokohama National University in 1994, 1996 and 1999, respectively. During 1999 to 2001 he was a Research Associate at Yokohama National University. Since 2001, he has been a assistant professor at Faculty of Environment and Information , Yokohama National University. His current research interests include the next generation database systems, multimedia databases, spatio-temporal modeling, query language and their implementation.

Tatsuo Tsuji received a doctoral degree in Information and Computer Science from Osaka University in 1978 and joined the Faculty of Engineering at Fukui University in Japan. Since 1992, he has been a Professor in the Information Science Department of the faculty. His current research interests include parallel and distributed database management systems. He is the author of the book, "Optimizing Schemes for Structured Programming Language Processors" published by Ellis Horwood. He is a member of IPSJ, IEICE, and IEEE.

Masahiko Tsukamoto, received his B.E., M.E., and Dr.E. degrees from Kyoto University, Kyoto, Japan, in 1987, 1989, and 1994, respectively. From 1989 to February 1995, he was a research engineer of Sharp Corporation. Since March 1995, he has been an Assistant Professor in the Department of Information Systems Engineering of Osaka University and since October 1996, he has been an Associate Professor at the same department. He is a member of seven learned societies, including ACM and IEEE. His current research interests include database systems, knowledge-base systems, and distributed computing systems.

Chiemi Watanabe, received her B.S. and M.S degrees in Information Science from Ochanomizu University, Tokyo, Japan, in 1998 and 2000, respectively. Currently, she is a doctor course student of the Graduate School of Humanities and Sciences, Ochanomizu University. Her research interests include spatio-temporal database systems, multi-modal computer human interactions, and information visualization.

Haruo Yokota received the B.E., M.E., and Dr.Eng. degrees from Tokyo Institute of Technology in 1980, 1982, and 1991, respectively. He joined Fujitsu Ltd. in 1982, and was a researcher at the Institute of New Generation Computer Technology (ICOT) from 1982 to 1986, and at Fujitsu Laboratories Ltd. from 1986 to 1992. From 1992 to 1998, he was an Associate Professor in Japan Advanced Institute of Science and Technology (JAIST). He is currently a Professor at Global Scientific Information and Computing Center and Department of Computer Science in Graduate School of Information Science and Engineering in Tokyo Institute of Technology. His research interests include parallel and distributed database engineering. He is a member of IPSJ, IEICE, JSAI, IEEE, IEEE-CS, ACM and ACM-SIGMOD.

Kazumasa Yokota received his B.S. in mathematics and Dr. Eng. degree in information science in 1972 and 1995, respectively, all from Kyoto University. Starting from 1985, he spent ten years at the Institute for New Generation Computer Technology (ICOT) engaged in the Fifth Generation Computer national project of Japan. He joined the faculty at Kyoto University in 1995 and at Okayama Prefectural University in 1997. He is now Professor in the Faculty of Computer Science and System Engineering.

Masatoshi Yoshikawa received his B.E., M.E. and Dr.Eng. degrees in information science in 1980, 1982 and 1985, respectively, all from Kyoto University. From 1985 to 1993, he was with Kyoto Sangyo University as a faculty member. In 1993, he joined Nara Institute of Science and Technology, and currently is an Associate Professor in Graduate School of Information Science. He is the chairperson of the Technical Group on Data Engineering of IEICE (The Institute of Electronics, Information and Communication Engineers). His current research interests include XML databases, databases on the Web, and multimedia databases.

Takeshi Yoshizawa, received his B.E. in health science in 1985 from the University of Ryukyu and his M.E. degrees in health science in 1989 from the University of Tokyo. Starting from 1989, he spent twelve years at the IBM Japan Co.,Ltd. engaged in the many database system projects. Now he is Advisory IT Specialist of Database Systems.

1

A New Database Technology for Cyberspace Applications

Yoshifumi Masunaga

Department of Information Science, Ochanomizu University

Chiemi Watanabe

Ayumi Ohsugi

Kozue Satoh

Graduate School of Humanities and Sciences, Ochanomizu University

ABSTRACT

In this study, we introduce a virtual world database system called "VWDB," which is currently under development at Ochanomizu University with the aim of creating a new generation database system for the cyberspace age. The system is constructed by integrating a virtual reality system and an object-oriented database system. In this paper, a VWDB prototype system architecture, a VWDB database schema definition language, and a VWDB data manipulation language are described. In particular, the design and implementation of the VWDB multi-modal data manipulation language is introduced, since the VWDB provides a virtual world as the user's database interface. It is natural to support a multi-modal database language so that users can interact with the VWDB database by means of gesture and voice. An automata-based approach is described for designing and implementing the multi-modal language.

1 Introduction

Traditional database systems such as relational or object-oriented database systems are weak in terms of supporting new database applications such as those in cyberspace . For instance, a "virtual shopping mall" with a database function such as a query and update capability cannot be constructed using a relational database system because the SQL interface is not intended to account for cyberspace applications. On the other hand, virtual reality (VR) systems are used to develop a variety of advanced cyberspace applications. For instance, a virtual world and a (distributed) virtual environment can be synthesized by using a virtual world description language

	Relational DB	Object-oriented DB	VWDB
Theory Based	Set Theory	Object-oriented Language (Smalltalk)	Virtual World (in contrast to the real world)
Target Data	Business Data	Engineering Data	Cyberspace Objects
Representation of Objects	Tuples	(Composite) Objects	Three-dimensional Spatial Objects
User Interface	Command Language	Command Language	Multi-modal User Interface
Year in Practical Use	1980's	1990's	2000's

Table 1: RDB, OODB, and VWDB -A comparison-

such as VRML or a tool kit such as World Tool Kit. However, VR systems have no database functions.

The idea of integrating a VR system and a database system occurred quite naturally because it seemed that such an integrated system could inherit both the VR capability and the database capability so that it could provide a set of all necessary functions to describe a variety of database-based cyberspace applications. We call this a "Virtual World Database System" (VWDB in short) . By using a VWDB, a customer can construct a three-dimensional (3-D) virtual shopping mall with database functions so that the customer can view merchandise such as television sets and refrigerators; moreover, the customer can ask about the price by voice, determine how to operate a door by hand, or even see through structure.

Table 1 depicts various features of the VWDB compared with those of the two traditional database systems, i.e., the relational database systems and the object-oriented database systems. This table also suggests future areas of inquiry that should be pursued in order that a virtual world database system (VWDB) may be realized. The differences are clear, and we believe that the VWDB will be widely used in the 21st century to develop advanced database applications.

The VWDB project was launched in April 1999 and was influenced by some of the previous work done in the "Block World Database System" (BWDB) development [1, 2] . As for related works, we would like to point out the study conducted by Tanaka et al. [3] in which a realization of LoD (level of detail)-based query capability in a VRML space was reproted, and the study conducted by Arisawa et al. [4] in which a spatio-temporal query capability in 3-D video databases was realized for the ergonomic analysis of human motion. However, to date, there has been no research reported regarding the integration of a VR system and a database system.

The rest of this paper is described as follows. The fundamental architecture of the VWDB is described in section 2. In section 3, we give an overview of a VWDB schema definition language, and in section 4 we describe how to design and implement a VWDB multi-modal database language. Section 5 concludes this paper.

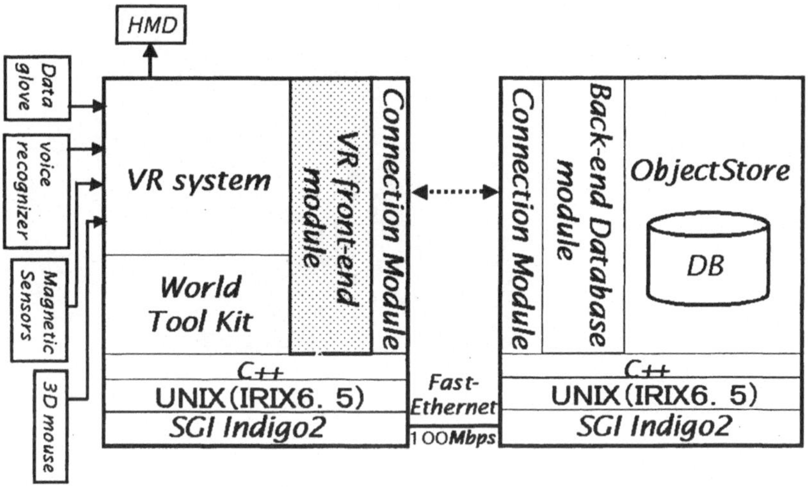

Figure 1: The current VWDB system architecture

2 The VWDB System

2.1 Prototyping

The VWDB is currently being developed at Ochanomizu University by integrating a VR system and an object-oriented database system. An object-oriented database system was selected because objects in a virtual world are usually composite objects and object- oriented database systems are much more efficient than relational database systems for storing and manipulating composite objects. Another reason for using an object-oriented database is that the message-passing paradigm of the object orientation seems suitable for supporting the multi-modality that will be a feature of user interactions with the VWDB. Figure 1 shows the current system architecture of the VWDB, in which a VR system and an object-oriented database system are connected via the Fast Ethernet. The architecture is scalable in the sense that more than one VR system could be connected to an object- oriented database system via a network, which would be necessary in order to realize a distributed virtual environment (DVE) for collaborative work support.

Figure 2 shows a set of input and output devices for users to wear when using the VWDB prototype system. By wearing an HMD (Figure 2-1), users can see the virtual world in 3-D. The 3-D 6DOF mouse (Figure 2-2) enables users to move from their current position in any of six directions, i.e., up, down, right, left, forward, and backward, in the virtual world. A magnetic sensor (Figure 2-3) is attached to the top of a user's head to capture the direction of the user's head as well as to detect the user's or the avatar's position when processing location-sensitive queries

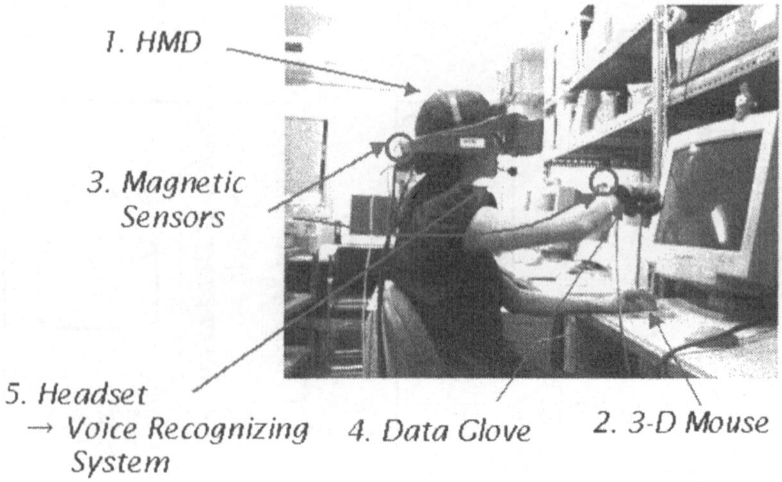

Figure 2: A user sitting at the VWDB prototype

and updates. These devices enable a user to walk through the virtual world as if he/she were in the real world. In addition to wearing the HMD and sensor and using the mouse, the user wears a data glove (Figure 2- 4) on his/her right hand and a headset on his/her head (Figure 2-5). The headset is connected to DS200, a voice recognition system from Speech Systems Inc. that processes a voice input based on the sentence pattern analysis method. These devices are necessary for measuring and recognizing the user's gestures and voice, and they ultimately allow the user to update and query the virtual world and its objects in a multi-modal manner.

Figure 3 shows a snapshot of a screen image of the VWDB prototype under development at Ochanomizu University. A virtual office with database function is prototyped. Users can become immersed in the virtual office, where the user's right hand is seen as his/her avatar. In the virtual office, it is possible, by means of multi-modal interactions, to move chairs and tables. A more detailed discussion of this is given in Section 4. Notice that, in the VWDB, a change in the virtual world is immediately reflected in the backend database, which is managed by Object Store, a commercial object-oriented database management system.

3 The VWDB Schema Definition Language: An Overview

3.1 Design Principles

To define and create a VWDB database, a schema definition language is necessary for VWDB enterprise administrators. We call it the VWDB Schema Definition Language. This language is different from a traditional virtual world definition language

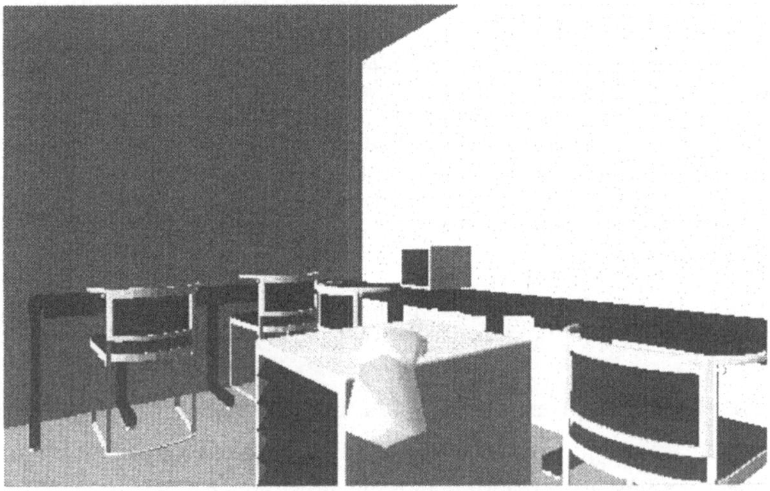

Figure 3: A screen image of a VWDB prototype

provided by VR systems in the sense that the former should describe a schema level definition of a virtual world, while the latter is enough to describe its instance level definition. For example, a "scene graph," which is known in VR to describe a virtual world, is an instance level description of the virtual world, yet it is not a schema level description of the virtual world. The schema level information is necessary because database systems use it to process queries and updates efficiently, i.e., to avoid exhaustive searches of databases to find targeted objects.

Another point we should consider in designing a schema definition language is that the VWDB objects are different from the objects in traditional object-oriented database systems in the sense that the former are usually "three-dimensional spatial objects" while the latter are not. Actually, a VWDB object has two different kinds of attributes as follows:

1. Spatial attributes such as the position, direction, and orientation of a VWDB object in a VWDB virtual database space.

2. Non-spatial attributes such as name, weight, color, texture, and so on.

Notice that, corresponding to the two types of attributes, two different types of domains are defined, and these are spatial domains and non-spatial domains, respectively.

Another feature of the VWDB database is that a set of objects in a VWDB database conforms to a certain class hierarchy that is well defined in the object-oriented paradigm. In VWDB, we call a set of similar objects a "category". In VWDB, objects belonging to the same category have the same spatial and non-spatial attributes as well as the same behaviors. Category hierarchies are defined similarly to the class hierarchies in the object- oriented paradigm.

Now, it becomes clear that the VWDB database schema definition language should have the following three sub-languages:

1. A virtual world definition sub-language for an enterprise administrator to describe a virtual world on the meta-level.
2. A category definition sub-language for an enterprise administrator to describe a category hierarchy.
3. A spatial domain definition sub-language to describe a set of spatial domain definitions to be used in the category definition above.

3.2 A Brief Description of the VWDB Schema Definition Language

In this section, we briefly show the syntax of the VWDB virtual world definition sub- language, the VWDB category definition sub-language, and the VWDB spatial domain definition sub-language.

3.2.1 The VWDB Virtual World Definition Sub-Language

The following shows a basic syntax of the VWDB virtual world definition sub- language:

```
World <Virtual World Name>{
  <Category Name-1>
  <Category Name-2>
       ...
  <Category Name-n>
};
```

In <Virtual World Name>, a VWDB database administrator or designer specifies the name of a virtual world that he/she wants to define. The designer specifies a whole set of category names to be used in the virtual world. Precise definitions of the categories are given using the VWDB category definition sub-language.

3.2.2 The VWDB Category Definition Sub-language

Using this sub-language, categories are defined in the following manner:

```
Category <category name>:<super category name>
{
  /*attributes*/
  <domain name><attribute name>;
  /*parts*/
  [<part category name><attribute name>;]
  /*material domain*/
  [<material domain mane>;]
  /*behavior*/
  [<behavior>;]
};
```

An example of a category definition is given below, where a category of OA desks is defined as a sub-category of category Desk. Note that an OA desk is a composite object that consists of two parts, i.e., a topboard and legs which are defined in category OA_desk_topboard and OA_desk_legs, respectively. As an example of behaviors, a constructor is shown here, where the initial values of the topboard, the legs, and the material of the legs are set using functions such as setvalue() and setmaterial value(), respectively.

```
Category OA_desk: Desk
{
  /*attributes*/
  char Owner[256];
  /*parts*/
  OA_desk_topboard topboard;
  OA_desk_legs legs;
  /*material domain*/
  /*behavior*/
  OA_desk(): Desk(), topboard(), legs(){
    topboard.setvalue(position, "0.0, 20.0, 0.0");
    legs.setvalue(position, "0.0, -20.0, 0.0");
    legs.setmaterialvalue(color, "0.6, 0.6, 0.6");
}
};
```

3.2.3 The VWDB Spatial Domain Definition Sub-Language

Spatial attributes need spatial domains to take on 3-D shapes as their values. In general, 3-D shapes are created using a 3-D computer graphics software or CAD software. However, such software describes a shape only as an instance. In other words, they do not provide a schema-level description of 3-D shapes. One exception is that the function "PROTO" defined in VRML allows a schema-level description of a 3-D shape in the sense that an arbitrary 3-D shape can be created by specifying a certain set of parameter values of a specified data type. For example, the next statement shows how to define a domain of 3-D shapes like the tables shown in Figure 4. The field statements specify a set of default values for the shape parameters. To specify a desired shape of the "ta007" PROTO type, a certain set of parameter values are specified in the constructor execution phase when a VWDB object is created or updated.

4 The VWDB Multi-modal Database Language

4.1 Design Principles

In the VWDB, it seems natural to realize a multi-modal user interface for VWDB users because users immerse themselves in the 3-D database space that is a virtual synthesized world by wearing and using various pieces of equipment, such as an HMD, a data glove, a 3-D mouse, a microphone, and sensors. To design a multi-modal database language for the VWDB, we first investigated an entire multi-modal

```
PROTO ta007
[ field SFVec3f VWtranslation 0 0 0
   field SFRotation VWrotation 1 0 0 0
   field SFVec3f VWscale 1 1 1
   field SFVec3f _X1_X2_ashiposition 0 0 0
   field SFVec3f _X1_X2_ashirotation 1 0 0 0
   field SFVec3f _X1_X2_baseposition 0 0 0
   field SFVec3f _X1_X2_baserotation 1 0 0 0
   field SFVec3f _X1_X2_tenbanposition 0 0 0
   field SFVec3f _X1_X2_tenbanrotation 1 0 0 0
   field SFColor _X1_X2_ashiColor 1.0 1.0 0.7812
   field SFColor _X1_X2_baseColor 1.0 0.9882 0.5859
   field SFColor _X1_X2_tenbanColor 0.8281 1.0 0.5312]
{
Transform {
   translation IS VWtranslation
   rotation IS VWrotation
      children [
         DEF _X1_X2_ashi Transform{
            translation IS _X1_X2_ashi_translation
            rotation IS _X1_X2_ashi_rotation
               Shape {
                  appearance Appearance {
                     material Material {
                        diffuseColor IS _X1_X2_ashiColor
                        ...
                     }
                  }
                  geometry IndexedFaceSet {
                     coord Coordinate {
                        point [ -78.721206 -32.5 -24.891851,
                        -78.947077 -32.5 -24.846922, ... ]
                        coordIndex [ 0, 1, 2, -1,3, 4, 5, -1, ... ]
                     }
                  }
               }
         }
         DEF _X1_X2_base Transform{
            translation IS _X1_X2_base_translation
            rotation IS _X1_X2_base_rotation
               Shape { ... }
         }
         DEF _X1_X2_tenban Transform{
            translation IS _X1_X2_tenban_translation
            rotation IS _X1_X2_tenban_rotation
               Shape { ... }
         }
      ]
}
}
```

Figure 4: A shape created by PROTO type ta007

interaction scheme in the VWDB, where the use of gestures and voice interactions are assumed. Since moving an object (hereafter called "the object move") is one of the most important operations in the VWDB, we studied this move intensively. Of course, beside the object move, instance object creation and deletion, composite object creation and deletion, and querying and updating non-spatial attribute values of an object will have to be considered to make the language complete, but these are left as future work. We must point out here the pioneering work in the field of multi-modal instruction with a computer system being done by Bolt [5], as well as the work by Tijerino et al. [6] in the area of verbal interaction of 3-D shapes in a virtual world. However, the commands being studied are not targeted database interactions.

The two main user modalities in a multi-modal VWDB are gestures made with the fingers of a data glove and the user's voice. Those two modalities will be combined so that a more complex and natural interaction can be specified. In designing a VWDB multi-modal database language, we noticed that an interaction initiated by a gesture or a voice instruction causes a "state change" of the target object, which is due to the change of the relationship between the user and the target object in the virtual world database. For example, suppose that a user starts to move a "chair" object, as shown in Figure 5. At first, the target object, i.e., the chair, is in the "initial" state because the relationship between the user and the object is nonexistent at this stage. Then the user may grasp the object in order to move it. At this point the state of the object may change from the initial state to the "ready-to-move" state. If the user's hand starts to move to the destination, then the state of the object is in the "in-moving" state. Now suppose that the user releases the object after reaching the desired destination. In that case, the state of the moved object may change to the "final" state because the move is finished. Since the object in the final state can accept the next interaction from other users, we may set the "final" state to equal the "initial" state via null input. Therefore the three stages for an object move are identified as follows:

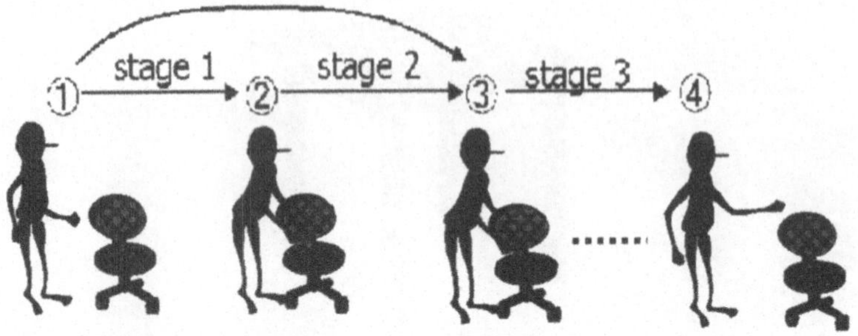

Figure 5. Relationship between a user and an object in an object move

Stage 1: The stage that transfers objects in the initial state to the ready-to-move state or even to the in-moving stage.

Stage 2: The stage that transfers objects in the ready-to-move state to the in-moving state.

Stage 3: The stage that transfers objects in the in-moving state to the final state.

4.2 Introduction of Automata to Characterize the Object Move

Let us try to characterize the state transition of an object move using automata. Stage 1 represents the state transition of an object in the initial state to the ready-to-move state or even to the in-moving state. This state transition can be done either by a sequence of voice instructions or by a sequence of gesture instructions in one of the following cases:

1. A user interacts with the virtual world by voice to identify the set of objects to move. For example, a user might utter a query such as, "Select all chairs whose weight is less than 10."
2. A user specifies a set of objects by gesture. This case is sub-divided into two cases:
   (a) Specify a set of objects by touching them with hand(s).
   (b) Specify a set of objects by pointing at them with hand(s).

Corresponding to cases (1), (2)-(a), and (2)-(b), three automaton inputs v1, g1, and g2 are introduced. Every object in the initial state q0 changes its state to state q1 if it matches the search condition specified in the voice instruction (v1). In the case of a gesture, an object changes its state from q0 to q2 or q3 when it is touched (g1) or pointed at (g2), respectively. The object in state q1 changes its state to a ready-to-move state q4 when the user utters "this" (v2) or "these" (v3). The object in state q2 changes its state to q4 if the user utters "this" (v2), while it changes its

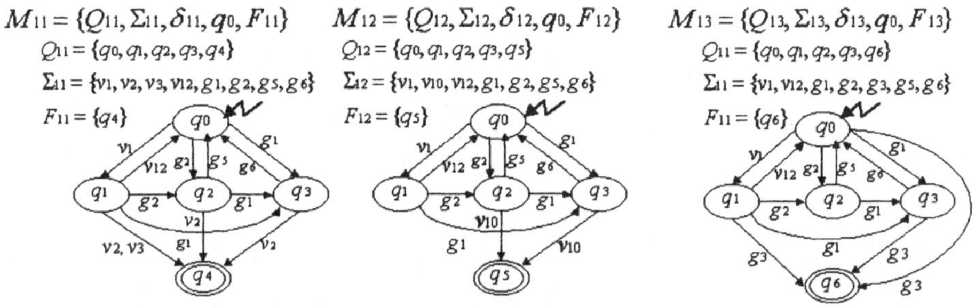

Figure 6. Automata characterizing stage 1

state to state q3 if the user touches the object (g1). The object in state q3 changes its state to state q4 if the user utters "this" (v2). The automaton M11 represents this state transition.

On the other hand, if the user utters "move" (v10) to the object in state q2 or q3, then its state changes to state q5. This state represents that the object is moving in a straight line toward the destination until a stop instruction is issued. This state transition is depicted by automaton M12.

If a user "grasps" (g3) an object in state q1 or q3, then it changes its state to state q6. This is the ready-to-move state, which is awaiting a subsequent instruction g7, where g7 represents the user's action to move the grasped object by moving his/her hand. Since some user might grasp an object in the initial state suddenly to move it, we allow state transition from q0 to q6 by g3, which corresponds to the direct arrow from node 1 to 3 in Figure 5. Automaton M13 represents this situation. Automata characterizing stage 1 are shown in Figure 6.

By similar argument, stage 2 and stage 3 are also characterized based on automata theory. To characterize an entire object move, these nine automata are connected using a method of cascading automata. As a result, the entire diagram is shown in Figure 7. In this figure, a sample multi-modal user interaction is indicated by a thick line going from q0 to q16 via q2, q4, q11, q8, and q5. First, a user pointed his/her finger at an object (g2). Then, the user uttered "this" to specify that this object was the correct object to move (v2). Next, the user pointed his/her finger in the desired direction to move (g2), followed by the user uttering "there" to identify the direction (v5), "move" (v10) to initiate motion, and finally "stop" to finish the move (v11).

4.3 Multi-Modal Input Translator

As was shown in the previous section, a sequence of multi-modal inputs that cause a state transition from the initial state (q0) to the final state (q16) is accepted as a multi- modal interaction for an object move in the VWDB. Since the VWDB prototype consists of a front-end VR system and a back-end object-oriented database

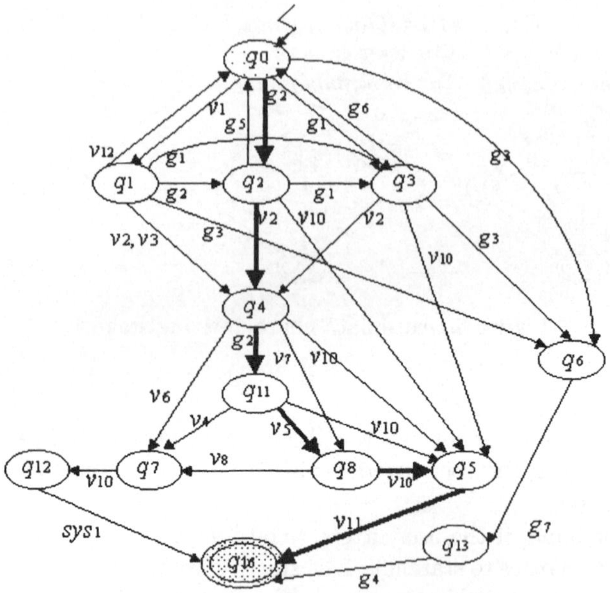

Figure 7. Automaton M characterizing an entire multi-modal interaction for an object move in the VWDB

system, a multi-modal input sequence accepted in the front end must be translated into a functionally equivalent "message" sequence to the back-end object-oriented system. In order to realize this translation, we have designed and implemented a module called the "multi-modal input translator," which is shown in Figure 8. The multi-modal input translator module consists of two sub-modules, the multi-modal input recognition module and the multi-modal input analysis and message generation module.

In order to illustrate how the multi-modal input translator works, we describe how the sample multi-modal interaction shown in Figure 7 is processed by these modules: First, a user pointed his/her finger at an object (g2), and then the user uttered "this" to specify that the object was the one he/she wished to move (v2). Next, the user uttered "forward" to identify the direction (v7), then "move" (v10), and finally "stop" to finish the move (v11). Now, the first module, i.e., the multi-modal input recognition module, outputs an alphabetic sequence "g2v2v7v10v12" with parameters, which is fed to the second module, i.e., the multi-modal input analysis and message generation module. Then, the second module interprets its syntax by constructing a parse tree, and generates a message sequence to the back-end object-oriented database system for database update. The message sequence is generated using a semantic network which is defined by the parse tree. More detailed

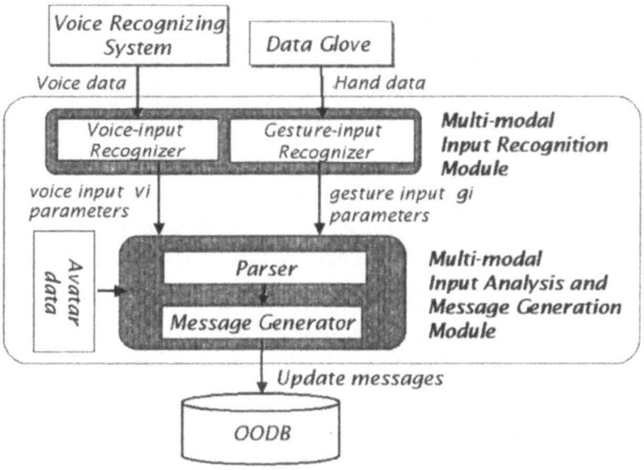

Figure 8. Data flow of the multi-modal input translator

description about the semantic network and its use is given in our recent work [7].

5 Conclusion

In this paper, we describe the design and implementation of VWDB which has been under development at Ochanomizu University since 1999. A prototype system architecture, a VWDB data base schema definition language, and a VWDB multi-modal data manipulation language were introduced in this paper. Particularly, the design and implementation of the VWDB multi-modal data manipulation language was intensively reported. Since users are wearing an HMD, a data globe, a 3-D mouse, a microphone, and sensors to detect the position of an avatar and a hand, users can interact with the virtual world database in a multi-modal manner. Two input modalities were voice and gesture. An automata-based approach was presented in this paper to characterize the VWDB multi- modal user interface.

The development of the VWDB has not been completed. We must expand our investigation to include a multi-modal query language as well as a virtual world constraint description language. Also, we need to expand our system to integrate more than one VR front-end systems so that the VWDB supports collaborative work.

Acknowledgements

The authors express their great thanks to all of the VWDB project members including Misses. Kumi Yoshikawa and Shoko Yoshida.

Bibliography

1) Masunaga, Y.: "The Block-World Data Model for a Collaborative Virtual Environment," Proceedings of the Second International Conference on Worldwide Computing and Its Applications (WWCA'98), Tsukuba, Japan, LNCS 1368, pp.309-324, Springer, March 1998.

2) Masunaga, Y., H. Kawashima, and Y. Mizuno: "Prototyping of the Block-World Database System," Transactions of Information Processing Society of Japan: Databases, Vol.40, No. SIG3 (TOD1), pp.91-104, February 1999 (in Japanese).

3) Kamiura, M., H. Oiso, K. Tajima, and K. Tanaka: "Spatial Views and LOD-Bases Access Control in VRML-object Databases," Worldwide Computing and Its Applications (WWCA'97), Tsukuba, Japan, LNCS 1274, pp.210-225, Springer, March 1997.

4) Tomii T., K. Salev, S. Imai, H. Arisawa: "Human Modelling and Design of Spatio-Temporal Queries on 3D Video Database," Fourth Working Conference on Visual Database System 4 (VDB4), pp.317-336, 1998.

5) Bolt, R. A. : "Put-That-There : Voice and Gesture at the Graphics Interface," Computer Graphics, Vol. 14, pp. 262 - 270, 1980.

6) Tijerino, Y. A., S. Abe, T. Miyasato and F. Kishino: "What You Say Is What You See - Interactive Generation, Manipulation and Modification of 3-D Shapes Based on Verbal Destinations," Artifical Intelligence Review, Vol. 8, pp. 215 - 234, 1994.

7) Masunaga, Y. and C. Watanabe: "The Design and Implementation of a Multi-modal User Interface of the Virtual World Database System (VWDB)," Proceedings of the Seventh International Conference on Database Systems for Advanced Applications (DASFAA 2001), pp.294-301, IEEE CS Press, Hong Kong, April 2001."

2

A Spatial Data Management Method for Integrating the Real Space with Virtual Space

Masahiko Tsukamoto

Department of Information Systems Engineering
Graduate School of Engineering, Osaka University

Takefumi Ogawa

Infomedia Education Division
Cybermedia Center, Osaka University

Shojiro Nishio

Department of Information Systems Engineering
Graduate School of Engineering, Osaka University

ABSTRACT

Due to advancements in technologies for virtual space construction, it has become possible for many applications to use a virtual space with high reality. Such a virtual space enables us to simulate a real space, and communicate with people separated at a long distance through networks. Several computing environments which integrate a real space with a virtual space have been proposed to realize communication among people in both spaces. In order to realize such an environment, it is necessary to manage a large amount of data in the real and the virtual space. To satisfy this requirement, we have been working on a fine-grained spatial data management method. We show how our method can manage fine-grained spatial data efficiently.

1 Introduction

With new virtual reality and computer graphics technologies, a virtual space can be constructed to appear as real as a real space. Accordingly, there are many studies of employing virtual spaces to support human communication and remote working. For example, FreeWalk [3)] realizes human communication by using virtual spaces constructed on a computer. Other approaches, such as the invisible person system [6)], introduce mixed environments, i.e., people in both the real space and the virtual space can communicate with each other by using the virtual space that models the real space. In many virtual spaces, people can go freely around and immerse themselves in the virtual space.

Most of conventional systems that manage virtual spaces process them as a whole. In such systems, virtual space is described based on a single global coordinate system. Basically, the system cannot handle anything that happens in the virtual space without all data of the space. As a result, it becomes hard to construct, manage and process a virtual space if it becomes large. In systems that mix a real space and a virtual space such as the invisible person system, it is often more difficult to construct, manage and process a virtual space. The most effective way to address such a scalability problem is to localize the description, the process, and the management of a virtual space.

In this chapter, we explain a method for integrating a real space with a virtual space based on fine-grained spatial data. The remainder of this chapter is organized as follows. In section 2, we explain the integration of real space with virtual space. The spatial data model is explained in section 3. Finally, we conclude this chapter in section 4 with some discussions about future work.

2 Integration of Real Space with Virtual Space

By using recent network technologies, we can easily build various kinds of virtual spaces that are accessible via the Internet. In such virtual spaces, a user who uses a home computer can communicate with other users in a similar situation even if they are separated by a long distance. They can make use of such an environment for disseminating, gathering, and exchanging several kinds of information. Moreover, the development of computer interface technologies such as virtual reality makes it possible for a computer user to achieve a wide diversity of social activities with more reality.

In general, activities in a virtual space do not directly affect the real space, and it has been pointed out that users hardly know the direct effect of their activities in the virtual space onto the real space. For example, when a user wants to buy something on a Web page, it is generally difficult for him/her to know the appropriate sequence of operations for fulfilling his/her intention. Such intentions could include the purchase of multiple goods at a time, the abortion of the current transaction, and the cancelling of a recent purchase. If the application system completely supports all users' intentions, users' operations may inevitably become complicated.

By integrating the real space with a virtual space, we expect to overcome many limitations in the conventional style of computing.

2.1 The 'Invisible Person' Approach

Based on the observation described above, we proposed a communication support environment called the 'invisible person' environment. In this subsection, we briefly explain the approach as an example of an integration of real and virtual spaces.

Figure 1: Invisible person environment.

Figure 1 shows the concept of the invisible person environment where a virtual space is constructed based on a part of the real space and the changes in the real space are instantly reflected onto the virtual space.

We call a person who visits a virtual space modeling of a part of the real space on a high-performance computer via a network an invisible person, because he/she cannot be seen by anyone in that part of the real space. In Figure 1, user A is an invisible person.

A person who is going to be an invisible person first runs a program on his/her computer. This computer is called an invisible person host (IPH). The program is similar to typical 3D browsers that are used to visit virtual malls and cyber cities, but it differs from them in one main point: the 3D space where the user is visiting is almost congruent with a part of the real space.

To grasp the changing situation in the real space that should be reflected onto the virtual space in real time, the system uses several kinds of sensors placed at various locations in the real space. Such sensors may include video cameras. The data taken by these sensors are analyzed to generate the associated virtual space. Note that the scene which each invisible person can see cannot be congruent with a complete real space, but should have reduced information. Here, when an invisible person visits a virtual place X, we call the situation "he/she is (virtually) at the (real) place X."

On the other hand, any person who really exists in the real space, called a real person, cannot directly see the persons who virtually visit there, and therefore we call them 'invisible' persons. In Figure 1, users B, C and D are real persons.

A real person can use a handy computer called an existence-sensitive pad (ESP) to see the invisible persons. An ESP is equipped with a small video camera that shows an image taken by the camera and the virtual images of the invisible persons visiting the place: it looks as if the invisible persons were there. Moreover, a conversation is realized when an ESP transmits the voice of each real person to the invisible persons and vice versa. We call a real person bringing an ESP an ESPer. In Figure 1, only user D is an ESPer.

An ESP may be equipped with a transparent head-mount display. In this case, the ESPer may see as if invisible persons were really in the real space.

The invisible person system is useful because we can realize human communication between a user in real space and a user in virtual space (i.e., a remote place). In particular, the realities of an invisible person and a real person provide each of them with a sense of existence and affinity of the other; this is very important in human communication. We show several useful situations in our daily lives.

A person who cannot go out from an office due to his/her tight schedule can be an invisible person and visit other places. He/she can meet other persons to negotiate business issues in his/her short spare time. He/she can also go home to do household matters. If he/she uses conventional tools such as email, a remote conference system, or remote operation tools, it is generally difficult to negotiate with other persons or to achieve complicated operations in his/her home. Therefore, in a practical sense, he/she has to go to the objective place to directly see the persons or to directly operate something. On the contrary, we may achieve most of such matters by using the invisible person system. If real persons can feel as if each invisible person exists in the place, the meeting or the negotiation will proceed smoothly. Moreover, if the virtual space reflects the real space in considerable detail, we will be able to achieve various kinds of complicated activities as an invisible person.

An employee who lives apart from other members of his/her family can have a dinner every night with them as an invisible person. A person who cannot take a long vacation can go abroad as an invisible person. Joining a telecommution or remote learning system as an invisible person will improve its effect compared to other systems. An invisible person has the chance to have occasional communication with those who are in a real place where he/she is currently visiting. A guard of a building, a construction foreperson, a sightseeing guide, and a participant of an auction have to be at certain places, and this can be effectively achieved if they become invisible persons.

2.2 The Design of a Prototype System

As for the realization of the invisible person system, it should be first noted that it is impossible for current computer technologies to realize a sufficiently realistic environment. There are many technical bottlenecks that prevent us from achieving, such as the concise recognition of real space, the user interfaces in virtual space, the immediate reflection of virtual space onto real space, and the real time processing of a huge amount of data.

In this subsection, we show our design of a prototype system [4)]. We focused on the following points:

- Reduction to a feasible design. We assume a common environment and do not use special or expensive hardware as much as possible. By using special hardware, we may enrich the reality of the environment, but the generality for easy deployment is lost.

Figure 2: Detection of a real person.

Figure 3: Display of management server.

Figure 4: An invisible person host.

Figure 5: IPH display.

Figure 6: An existence-sensitive pad.

Figure 7: ESP display.

- Wide-spread popularity to become an invisible person. A user can become an invisible person without using special hardware and software. Moreover, through the Internet, anyone can visit the environment from anywhere in the world as an invisible person and thus communicate with other persons in the world. Standard protocols and tools should be used for Internet access.
- Emphasis on the realization of communication rather than the concrete analysis and accurate presentation of the real space. Since facial expressions and gestures are important in human communication, we prefer to use the real time video images of users.

Based on this policy, we have designed the system in the following way. In the real space, we fix a graphic workstation and equip it with CCD cameras, microphones, speakers, and a wireless LAN interface. Figure 2 shows the situation when a real person is detected using a CCD camera. This computer is called the management server (Figure 3). The video image of real persons is taken by one or some of the CCD cameras, and a minimal bounding box of each real person is cut from them, which is then used to compose the virtual space seen by an invisible person, as shown in Figure 4. Here, we assume that objects, except for persons, do not move, and should be modeled by a system administrator beforehand. The management server communicates with all IPHs and ESPs. Several microphones are fixed at some locations in the real space to catch sound and transmit it to each IPH.

A microphone, a speaker, and a CCD camera are attached to an IPH as shown in Figure 5. The CCD camera is used to capture the users' facial image. The captured video images are transferred to the management server. The voice of each invisible person is captured via the microphone, and then transmitted to the management server. It is played by one of the speakers nearest to his/her imaginary location in the real space.

An ESP is a notebook personal computer equipped with a CCD camera, a wireless LAN card, and a sensor which obtains the location information of the ESP, as shown in Figure 6. The location information and the image data of each invisible person are sent by the management server via the wireless communication. Based on the obtained information, the ESP calculates the x/y coordinate values for each invisible person to paste its images in the ESP's display. Figure 7 shows some images taken by an ESP's CCD camera composed with the facial images of invisible persons.

3 Spatial Data Model

In order to integrate the real space with the virtual space, we should use a virtual space modeled on the real space to simulate many things which can occur in the real space. In this case, it is necessary to construct a large virtual space so that many avatars can be activated in the virtual space. Most of the conventional systems that manage virtual spaces process them as a whole. In such systems, the virtual space is described based on a single global coordinate system. Basically, the system cannot handle anything that happens in the virtual space without all data of the space. As a result, it becomes hard to construct, manage and process a virtual space if it becomes large. The most effective way of dealing with this scalability problem is to localize the description, the process and the management of a virtual space. We show our approach in this section.

3.1 Fine-grained Spatial Data Models

In order to localize the description, the process, and the management of a virtual space, we divide the virtual space into multiple subspaces. Each subspace is set as an independent coordinate system to maintain the independency of each subspace. Events/Actions which occur in the subspace are processed and managed based on the coordinate system of each subspace. We call the subspaces scenes.

A virtual space becomes scalable when it is divided into small scenes, i.e., it can be represented by a set of small descriptions of spaces that are tractable by a computer. However, the following three points should be considered:

- How to express data in a simple manner.
- How to simplify the topological association of each scene.
- How to make the management of each scene independent.

A typical approach to the first problem is to use the box shape of a subspace. If the user's activity in the space is essentially 2D, a rectangular or square shape of the floor can be used. Since we can easily put such boxes (or rectangles) side by side to construct larger regions of space, this approach seems to be effective to solve the second problem. In this case, the topological association will be defined by edge-sharing. Many associated regions may construct a grid topology, which means each region can be identified by the row and the column numbers. Although the grid topology is easy to manage, it does not solve the last problem. The global coordinate of the region arrangement should be defined a priori. The division and merging of regions cannot be done locally. Furthermore, two different spaces cannot be easily combined.

A possible solution to the last problem is the scene-based approach. That is, each region can be independently constructed without assuming any global view, and multiple regions are linked together to construct the local association, or topology. We call such a unit of region a scene. If the scene shape is rectangle, four links typically are defined where each link corresponds to each edge of a rectangle. This is different from the grid approach because it does not have any global constraints.

3.2 The IBNR Approach

When we visualize a scene in the scene-based approach, we may take several other scenes into consideration since we can see distant places if there is no obstacle. This fact worsens the scene independency. In order to maintain independency, we should restrict something in the visualization. In this section, we force a fixed viewpoint constraint into a scene visualization. We consider that a simple method for constructing complex spaces with reality is to use of a picture taken in the real space as the background. Based on this idea, we propose an approach called image-based non-rendering (IBNR) [7)]. We use this phrase as an antithesis to a recent VR trend, i.e., image-based rendering (IBR), where images are, in many cases, deeply analyzed and distorted to construct a 3D model. In contrast, images are used without any analysis and distortion in our IBNR approach.

3.2.1 The IBNR Data Model

From pictures taken in the real space, the IBNR system constructs a virtual space with high reality. The pictures make it easy for a creator to construct a virtual space that contains many objects in the real world without any modeling.

In IBNR, we first prepare a scenic image such as that shown in Figure 8 and eight human images such as those in Figure 9. In these human images, a person is directed to the front, back, right, left, and their intermediates. Next, as shown in Figure 10, we give the trapezoid of the scenic image to specify the floor region where a user can move by using an avatar. Here after, we sometimes use the word 'user' for the avatar.

The IBNR scene data consist of the floor information and link information as shown in Figure 10.

1. Floor information

Figure 8: A scenic image for background.

Figure 9: Human images for an avatar.

Figure 10: Data model of an IBNR scene.

floor region: A scene creator sets the bottom left, bottom right, top left, and top right coordinates of the trapezoid in the picture.

floor size: We assume that the shape of a floor is a rectangle and the rectangular floor in the real space corresponds to a trapezoid area in the picture. A scene creator specifies the width and the depth of the floor.

2. Link information

   The link information represents the linkage to other pages from the front, back, right, and left sides of the floor. He/she specifies a linking URL and a segment of the corresponding side in the trapezoid to each side. Four HTML files are prepared for one scene. These files are identified by the last characters of the file name, 'f', 'b', 'l', and 'r', which stand for 'front', 'back', 'left', and 'right', respectively. They represent the entry direction. That is, when the scene name is 'x0', there are four files 'x0f.html', 'x0b.html', 'x0l.html', and 'x0r.html', and they are different in the initial status of an avatar.

3.2.2 Visualization of a Scene

Based on this configuration, the system can construct a scene which a user can immerse himself/herself in and walk through. As shown in Figure 11(a), the picture in Figure 8 is used as the background and one of the pictures in Figure 9 is composed in an appropriate size. A person's picture is enlarged and shrunk by the user's input to represent the depth of the pseudo-3D space. In this figure, if the user presses a key (e.g., 't') assigned to the 'forward' operation several times, the avatar picture is gradually enlarged as shown in (b), as if the avatar steps forward in the space shown in the background picture. If a user wants to change the direction of the avatar,

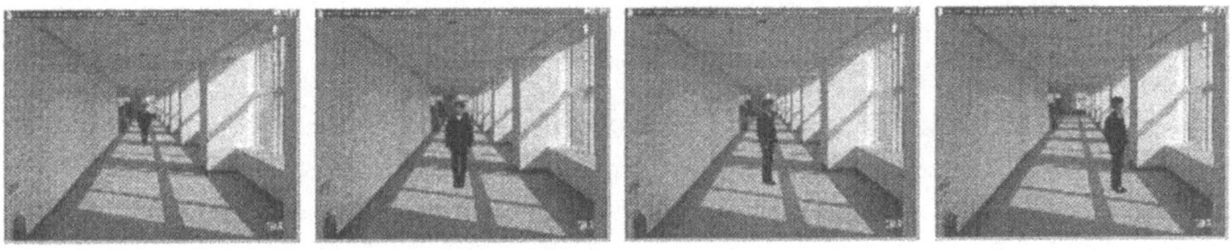

(a) (b) (c) (d)

Figure 11: Examples of image compositions.

Figure 12: Example of a new scene.

he/she can input another key (e.g., 'f') assigned to the 'rotate to left' operation by which the picture is changed as shown in (c). In addition, if the user inputs 't' several times to step forward, he/she obtains the scene shown in (d), which means the direction of the avatar's step is changed. An example of key configurations is 't' to go forward, 'f' to rotate to the left, 'h' to rotate to the right, and 'b' to go backward.

In Figure 11(a), if the user repeats going forward and the avatar comes out of a predefined floor region, i.e., the bottom side of the trapezoid shown in Figure 10, the scene automatically changes to another one, as shown in Figure 12. This figure where the view point and the direction of the scene are different from those of Figure 11(a). This scene is constructed in a similar manner to the scene shown in Figure 11, where the background image, the trapezoid region, and the linkage information are different. Note that, in the situation shown in Figure 12, if a user turns back and moves to the back, the scene again changes back to Figure 11(a), where the avatar appears with his backside image in front of the scene.

In each picture in Figures 11 and 12, there are two or three arrows at the right top area. These arrows represent the directions to which other scenes are linked from the scene. Upward, downward, leftward, and rightward arrows, respectively, link the scenes to the back, front, left, and right of the scene.

HTML, JavaScript, and Java are used to describe scenes that can be displayed on a normal WWW browser. In each scene, four URLs at most are attached to represent the front, back, right, and left neighbors of the scene. Of course, an attached URL is not necessarily an IBNR scene, i.e., it can be a normal WWW page. The normal pages are also effectively used in the entrance of a building, an office, and a room to put a separation into a sequence of scenes.

In general, there are many functions of WWW which are used for a variety of purposes. These functions can also be utilized in constructing IBNR contents. By using the cookie, we can construct a 3D content where a scene is changed if a user visits the place after visiting another certain place. Several interesting contents with scenarios are typically seen in RPG and sound novels. An example is as follows. A user cannot come into a building unless he/she visits a place and obtains a key. Another example is that the shutter is closed if a user turns on a switch in another scene.

We can use a page that plays a video. In IBNR contents of a zoo or an aquarium, if a user comes close to a cage or a watertank, a video showing animals in the cage or fish in the watertank is played. If an avatar goes into a train or a car, a video showing the avatar going in the vehicle and the vehicle exiting is played. Such dynamic contents can enrich the IBNR contents.

3.3 Extending IBNR System for Multi-user Communication

IBNR is an effective method for easily constructing realistic virtual spaces. If IBNR is a multi-user system [5)], it can be used by anyone as a communication system on the WWW like conventional ones using virtual spaces.

The communication space constructed with IBNR uses scenery photographs as background images. Therefore, for different types of communication, different virtual spaces can be chosen as described in the following examples:

- Scenes of restaurants, seasides or large grass fields may be a good choice for relaxed conversations.
- Scenes of beautiful seashore during sunset are suitable for conversations with loved ones in a romantic mood.
- Important discussions should be held in the conference room.
- It is appropriate to choose a place where few people pass by for secret discussions.

Thus, communication in a real atmosphere is made possible in IBNR scenes.

3.3.1 System Design

People communicate by exchanging information through verbal communication and nonverbal communication (gestures, voice tone, facial expression, eyes) [2)]. From this information, we can understand the interests of other participants regarding the topic of conversation and the atmosphere of the conversation, and predict whether someone nearby is joining the conversation. When there are many people in a small place, much information is available and one may get confused in understanding the information. This makes the communication process complex and it cannot be carried out smoothly. However, in the real world, even in a crowded train, communication is carried out smoothly. This is because human communication in the real world has some special features. In this subsection, we describe how these features are introduced to multi-user IBNR.

3.3.2 Communication in the Real World

In the real world, one can see the figures of others in the same room except when they are hidden by some obstacles; even when one cannot see others in the same room, one must be able to hear them. This is caused by the continuity feature of the real space, where light and sounds are transmitted. The effect of the transmitted information decreases with the increase in distance. A person standing farther away looks smaller and not so clear, and his/her voice may be hardly heard. Consequently, one usually filters out such kind of unimportant information from a distance. In other words, one usually moves closer to the information he/she needs, and keeps a distance from the information he/she does not need.

Following the Principles of the Cyberspace by Benedikt [1)], we proposed the above as a new principle called the Principle of the Information Transmitting (PIT), and regard it as an important factor in constructing virtual spaces.

3.3.3 Information Transmission in Multi-user IBNR

Multi-user IBNR allows the sharing of visual and verbal information only among users in the same scene to maintain the independency of each scene. Therefore, when one constructs a pseudo-3D space by dividing a room into several scenes, users accessing the same room in different scenes are not aware of each other. In other words, the communication space constructed in multi-user IBNR is a rather limited space in terms of communication compared to the real world.

To make the communication space provided by multi-user IBNR more flexible, we alter IBNR's independency feature to allow adjacent scenes to transmit information. By applying the PIT to the virtual communication space, users can filter out unwanted information.

3.3.4 Visualization of a Scene Based on The PIT

In multi-user IBNR, the existence of avatars and their conversations are spread to adjacent scenes based on the PIT.

The figure of the avatar: As the distance increases, the avatar becomes dimmer, and thus instead of being able to know who is out there, the user can only knows that someone is out there. This is implemented by showing a dark, unclear image of the avatar.

Conversations: Similarly, we apply the PIT to multi-user IBNR, and adjust the size of the balloon which shows the conversation and the font size of the text to

Figure 13: Example of login screen.

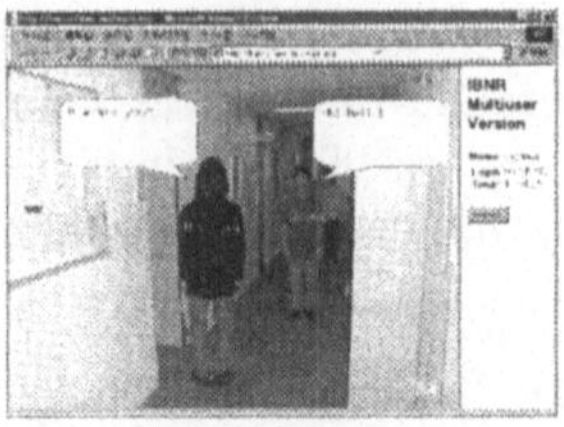

Figure 14: Example of main screen.

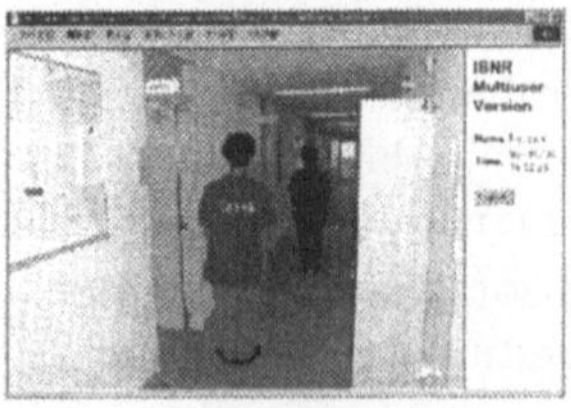

Figure 15: Avatar in the far scene.

reflect the volume of the voice. For example, when a conversation is held in an adjacent scene with a small voice, it can be reflected by using a small font size; when a conversation is held in a farther scene, the text is shown as dotted lines "..." to imply that someone is speaking in a small voice. By considering the intentional change of volume while talking, which is like calling someone or having a confidential conversation, multi-user IBNR allows users to set the volume of their "voice". When one talks in a normal tone in an adjacent scene, the text shown is smaller than that when he/she talks in the same scene. However, if he/she talks loudly, the text shown on the screen is of a normal size as if he/she speaks in the same scene. Conversely, when one speaks in a soft manner, the text is converted into dotted lines "...".

The figure of other avatars in nearby scenes can be seen. In such a way, a conversation group may be easily formed. If there are several groups engaged in conversation, one can easily judge which group to join by listening to the content of the dialogues.

In the login screen shown in Figure 13, the user can input his login name and select an avatar. Six avatars are prepared but the user can also use his own avatar image by using the text box at the bottom to input the URL of the avatar image. After logging into the system, the user can walk through the pseudo-3D space shown on the main screen (Figure 14). The maximum number of avatars is limited to 10 avatars per scene to avoid system performance degradation. Message input is done in the message-input window that appears when the user presses a certain key. The message is displayed in a balloon shown on each browser of all users in the same scene (as in Figure 14).

The balloon that contains the message is displayed near the avatar image. An avatar in a distant scene not visible from the current scene is displayed simply as a human figure (Figure 15). The human figure is only displayed to tell his/her existence. In terms of messages, the message from an avatar in the same scene is displayed with a normal size; on the other hand, the size of the message from an avatar in an adjacent scene is changed according to the volume of the avatar's voice, as shown in Figures 16, 17 and 18. Fig. 16(b) shows the adjacent scene of Figure 16(a) in the same situation as displayed on another user's screen. Scene (a) cannot be seen from (b), therefore only the balloon is displayed.

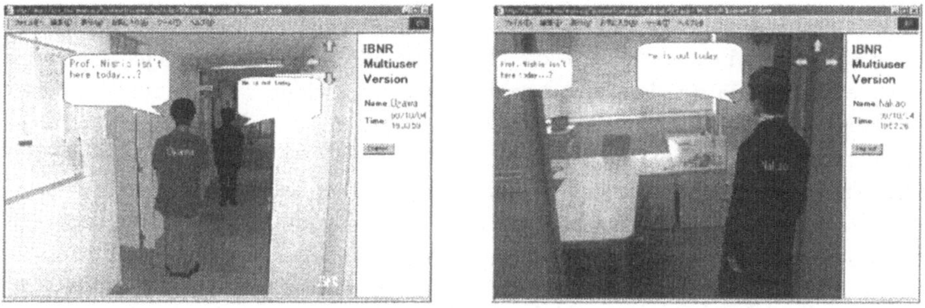

(a) (b)

Figure 16: Message in normal voice.

Figure 17: Message in loud voice.

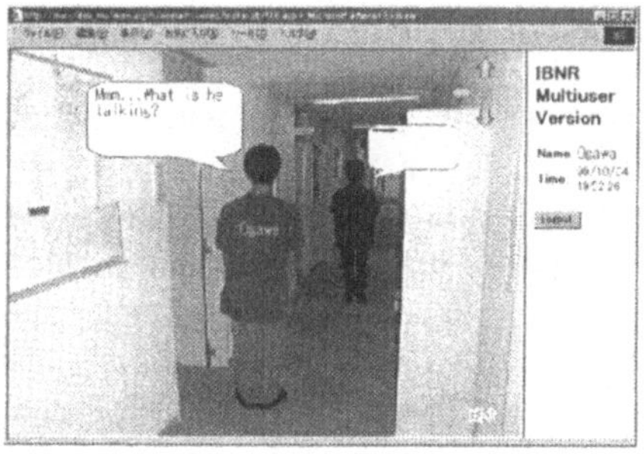

Figure 18: Message in weak voice.

3.4 Putting Real-time Real World Aspects in IBNR

In the first prototype of the invisible person system described in section 2, there are several serious problems. First, the construction of virtual space based on real space has a high cost. Second, visiting such a virtual space on WWW also takes a lot of transmission, rendering, and management, especially when it is very large or elaborated. Third, it is difficult to reflect the situation in the real space on the virtual space in real-time. The first two problems can be solved by introducing the IBNR multi-user version. Thus, a simple way to solve the problems is to extend IBNR to manage the third problem, i.e., to put real-time, real world aspects.

This can be easily done by using video images instead of scenic images. In this case, virtual space that simulates things in real time which occur in the real space can be constructed. At the implementation level, the simplest way for using real-time video in WWW is to use 'meta' tags for refreshing the page in a short period of time, possibly every few seconds. If more frequent refreshing is needed, several real-time video streaming technologies can be used, and they generally request plug-ins of WWW browsers. We have implemented the IBNR version of IPH for the invisible person system. It is constructed using the former method, and the real-time video is analyzed by the video server to extract real people in the video. The images of real people are pasted on the video image as avatars (which we call 'real' avatars) with appropriate depth information so that the depth among virtual avatars and real avatars is properly visualized in IBNR.

The aspects of independency in IBNR bring another merit to the operation of the invisible person system. This is a provision for the scalability problem: multiple servers can be used for different scenes. Therefore, even when a vast number of users access a certain region of a virtual space, the performance will not degrade if it is divided into many scenes. In this way, a lot of people can use the invisible person system in this version. We should note that we employ a normal VRML browser for remote people. Furthermore, we can link space constructed in the previous version with that in the IBNR version by using a normal WWW link.

4 Conclusion

We have described environments for integrating real spaces with virtual spaces. In order to realize such an environment, we must manage a large amount of unassembled data in the real space and the virtual space. From this viewpoint, the IBNR approach introduced here is scene-based; each scene can be constructed in a bottom-up manner independently from other scene constructions. This approach is rather restrictive, and a lot of work remains to be done to realize future environments where real space and virtual space are integrated at a more advanced level. It may be an extreme case from the 3D modeling point of view. Another extreme is full 3D models such as VRML, where an independent division of a whole space is quite difficult. In that sense, we need to integrate of these two approaches so that full 3D visualization is possible with independently constructed local data. This requires a more sophisticated data model, and an extension to IBNR can be a step toward this goal. Creating a more seamless linkage between two neighbouring scenes is also a problem.

The current scene change must be intermittent. Some kind of 3D-aware morphing technique is necessary to make the scene change continuous. Furthermore, we should consider the reusability issue, i.e., virtual space data should be reusable by implementing a VRML compiler tool for constructing IBNR scenes from data described with VRML.

Acknowledgements

We wish to thank Prof. Fumio Kishino, Prof. Yoshifumi Kitamura, and Prof. Masatoshi Arikawa for their valuable comments on this paper. We are deeply grateful to Mr. Hiroaki Hagino, Mr. Tsutomu Terada, Mr. Yutaka Sakane, and Mr. Satoshi Nakamura for their cooperation in creating the IBNR contents. We would like to thank other members of Nishio Laboratory for their helpful comments and suggestions on this research. This work is supported by Research for the Future Program of Japan Society for the Promotion of Science under the Project "Researches on Advanced Multimedia Contents Processing" (JSPS-RFTF97P00501).

Bibliography

1) M. Benedikt, Cyberspace: Some Proposals. In *Cyberspace: first steps*, M. Benedikt (ed.), The MIT Press, 1991.

2) T. Kurokawa, *Non-verbal Interface*. Human Communication Engineering Series, Ohmsha, Ltd. 1994, in Japanese.

3) H. Nakanishi, C. Yoshida, T. Nishimura, and T. Ishida, FreeWalk:Supporting Casual Meetings in a Network. In *Proc. of the ACM 1996 Conference on Computer Supported Cooperative Work*, ACM CSCW, 1996, pp. 308–314.

4) T. Ogawa, Y. Sakane, Y. Yanagisawa, M. Tsukamoto, and S. Nishio, Design and Implementation of a Communication Support System based on Projection of Real Space on Virtual Space. In *Proc. of 1997 IEEE Pacific Rim Conference on Communications*, Computers and Signal Processing, 1997, pp. 247–250.

5) T. Ogawa, M. Tsukamoto, and S. Nishio, Realizing Collaborative Pseudo-3D Space on WWW. In *Proc. of Data Engineering Work Shop (CD-ROM)*, IEICE 1999, in Japanese.

6) M. Tsukamoto, Integrating Real Space and Virtual Space in the 'Invisible Person' Communication Support System. In *Proc. of 1st International Conference on Advanced Multimedia Content Processing*, 1998, pp. 62–77.

7) M. Tsukamoto, Image-based Pseudo-3D Visualization of Real Space on WWW. In *Digital Cities: Experiences, Technologies and Future Perspectives*, T. Ishida and K. Isbister (eds.), Lecture Notes in Computer Science 1765, Springer-Verlag, 2000, pp. 288–302.

3

Database Support for Cooperative Work

Yusuke Yokota

Kyoto University

Yahiko Kambayashi

Kyoto University

ABSTRACT

Most of CSCW applications are implemented as network applications, which support cooperation among users in distributed network environment. Such applications have to provide facilities of sharing and modification of resources for collaboration. Databases are suitable for foundations of these facilities because of its characteristics: separation of data and applications, support of distributed environment, transaction management, and data integrity. To realize some requirements from CSCW systems, extension of database functions are needed. We introduce a cooperative hypermedia system VIEW Media, which is based on database technologies and behaves as a framework for CSCW systems. The system has features of database systems and realizes development of flexible cooperative work application.

1 Introduction

CSCW (Computer-Supported Cooperative Work) is a research area which pursues a way of collaboration of users through computer environment. Database technologies will support realization of robust, secure and flexible CSCW applications.

Most of CSCW applications are implemented as network applications, which support cooperation among users in distributed network environment. Such applications have to provide facilities of sharing and modification of resources for collaboration. Databases are suitable for foundations of these facilities because of its characteristics: separation of data and applications, support of distributed environment, transaction management, and data integrity.

Consideration of security and privacy is a important issue for CSCW applications to support practical cooperative work. Database technologies have a capability to introduce security mechanisms into CSCW applications. In addition to the conventional security mechanisms which database technologies provide, CSCW applications present a need for extension of the mechanisms coping with various situation of collaboration.

A view mechanism of databases is also important for CSCW applications to provide flexible workspaces for users. Conventional CSCW applications adopt WYSIWIS (What You See Is What I See) principle as a foundation of construction of systems. This principle assures that all users see the same contents of data. Thus it is useful to form common recognition among users, though limits flexibility of collaboration. To release CSCW applications from this limitation, introduction of the view mechanism and its extension are required.

The notion of awareness originated from CSCW researches. Awareness is understanding of the activities of others, which provides a context for user's own activity. When WYSIWIS principle is not always be effected, the provision of awareness information becomes more important and various methods of providing awareness information are required. Controllability and observability are generalization of awareness. These notions materialize more controlling and processing awareness information in detail. Controllability enables specific users to control other specific users by force, which is required when users who have various roles work cooperatively.

2 Database Functions Required by CSCW Systems

Generally, CSCW systems are implemented as distributed systems. Though database functions will help develop such CSCW systems, few researches have indicated the availability of the functions in the development of CSCW systems. In this section, database functions applicable to the development of CSCW systems are outlined.

2.1 Data Sharing among Users and Applications

There are two major approaches to construct CSCW systems: converting single-user applications into multiple-user ones and developing collaborative applications utilizing a toolkit designed for such purposes. The former approach usually provides developers with a framework to support the conversion and realizes application sharing. Although it enables developers to reuse applications designed for single-user use and reduces the costs of developing collaborative applications, it is difficult to implement various functions for supporting collaboration. The latter approach is rather straightforward. Functions for supporting collaboration will be realized easily in this approach.

Data sharing among users is a basic requirement of CSCW systems. The feature will be supported by both approaches. Though less attention is paid to data sharing among applications in conventional CSCW systems, this type of sharing is important from the point of view of reuse of data. To realize data sharing among applications, separation of applications and data is required. Application conversion approach is not suitable to support both types of data sharing, because source single-user applications usually have no consideration for the separation. A toolkit for collaborative applications supporting both types of data sharing is required to develop CSCW systems which support large scale and long time collaboration.

2.2 Persistency and Reuse of Data

In general, collaboration on CSCW systems classified into two types, synchronous and asynchronous collaboration. To support asynchronous collaboration, persistency of data is indispensable. Reuse of data is important to CSCW systems, because it enables users to share data as results of collaboration and extends lifetime of data. Furthermore, persistency of data bridges synchronous and asynchronous collaboration by application of action history and action history view functions[21)]. These functions store the sequence of operations of users into a database. Users are able to replay it later selecting part of the sequence data.

2.3 Customization of Data

To construct practical CSCW systems, the capability to customize data is important as well as to share data. Generally, there are different roles or characters of users and different kinds of computers users use in cooperative work. These differences require the functions for customization of data. It realizes that users can obtain different views of the same data according to the difference of their roles or characters, and computers. For example, in a distance learning system, a teacher and students share the same textbook data. The teacher may want to add memos about teaching hints and the students may want to annotate it for a reminder of a topic. Heterogeneous computing environment also requires customization. A screen of a PDA can not provide the same capability of a desktop PC. When a user of a PDA sees some document, a system has to simplify the document to display it on a screen of the PDA.

In CSCW systems, two types of customization are required: customization of representation of data and customization of contents or structure of data. Customization of representation of data does not change meaning of data. It enables users to change window size, font size, types, and color. Users can also see different parts of the same document, which is called location relaxed WYSIWIS principle[23)]. WYSIWIS (What You See Is What I See) principle is a concept of CSCW systems that users in cooperative work see the same representation of the same data. It enables users to share the same recognition related to the cooperative work in progress and helps users to promote work smoothly. However, the principle does not suit all situations of cooperative work and the methods that relax the strict WYSIWIS principle are required to support more flexible collaboration. In contrast, customization of contents or structure of data changes meaning of data: customization of contents modifies internal elements of atomic data components and customization of structure modifies relationships among data components. This type of customization is used when users intend to customize data for their own purposes or users are not permitted to access some data because of security reasons. For example, users can hide part of documents or add more contents as parts of documents. This operation should not affect source documents. Many CSCW systems lack such functions.

The contents of customization are important as results of collaboration. They represent the knowledge of users that is produced during collaboration. Thus, the function of sharing customization is needed to share the produced knowledge among users. This function provides an opportunity to integrate the notions of customization and authoring in cooperative work. Authoring is to create or modify shared data and the results of the work are always shared by all users. Though customization is to create or modify shared data and the results of the work are not shared by other users basically, the function can control the sharing level of data. The contents of shared customization that has maximum level of sharing will be equal to that of authoring.

2.4 Active Capability of a System

Active capability is required to develop various event-driven services on CSCW systems. ECA rules are suitable for the foundation of such services.

Updating graphical components on a screen is a typical requirement of synchronous CSCW systems. Most of graphical components represent contents of shared data or statuses of elements of a system, namely, these components are dependent on shared data or elements of a system. Thus these components should be updated when their source data or elements are changed to reflect the newest contents or statuses on a screen. An event filtering mechanism which controls unnecessary or excess events is important in reducing communication costs.

COAST extends MVC architecture for distributed environment. MVC architecture also utilizes an event-driven mechanism for notification of changes on models to their views. The architecture separates a component into three parts: a model as a container of data, a view as representation of data, and a controller as a channel of communication between a system and a user. The distributed MVC architecture enables users to share the same model among users by multiple views distributed over the network.

Notification services are more application specific requirements. Users in collaboration may want to be notified when statuses of certain components of the system (including data and other users) are changed.

Collaboration on some kinds of cooperative applications progresses according to a change of status of work. These kinds of applications are able to adopt the ECA rules to describe the logic of the applications.

To meet these requirements, an ECA-rule mechanism for CSCW systems need to be implemented as a network-transparent mechanism: event detection, condition checking, and action performing in one rule will be executed on separate hosts. For example, when a CSCW system which has client-server architecture utilizes the distributed ECA-rule mechanism, event detection and condition checking are executed on a server side and an action is performed at a client side.

2.5 Transaction Management

CSCW systems have to support both short transactions and long transactions, because various types of cooperative work of users require both of them. One of the typical solution to support long transactions is to provide two types of workspaces (or databases), public and private ones, with check-in and check-out mechanisms. A user can import data from a public workspace into her/his own private workspace by checking out. When the user finished modification of the data, s/he can return the modified data to the public workspace by checking in. Though most of the existing CSCW systems support the two types of workspaces, more flexible solutions are expected to support more extensive cooperative work.

2.6 Security and Privacy

To support practical cooperative work, CSCW systems need to provide mechanisms protecting security and privacy of users. There is a trade-off between protection of security and privacy and flexibility in sharing and collaboration in CSCW systems. Thus, security mechanisms of CSCW systems should offer a dynamic access control framework to users so that they can change security levels during collaboration.

3 VIEW Media: a Framework for CSCW systems

VIEW Media is a framework for CSCW systems developed by our laboratory. It adopts database technology as a foundation of its architecture. VIEW Media is designed as a collaborative hypermedia system. Developers can build applications for specific purposes on the system, such as a distance education system, a distributed conference system, and so forth.

3.1 The Fundamental Objects of VIEW Media

VIEW Media defines four types of fundamental objects constructing workspace of the system:

- *User surrogate objects* represent users participating in workspace. A user surrogate object contains profile and state information of a corresponding user.
- *Hypermedia component objects* organize shared hypermedia documents. Hypermedia is used as a basic data format for cooperative work on VIEW Media. Results of collaboration are stored as hypermedia component objects in an object repository of VIEW Media. The hypermedia model of VIEW Media was designed for flexible customization, which is based on the Dexter Hypertext Reference Model[28)].
- *Equipment objects* provide various communication methods for interaction among users such as chat tools, shared pointers and so forth.

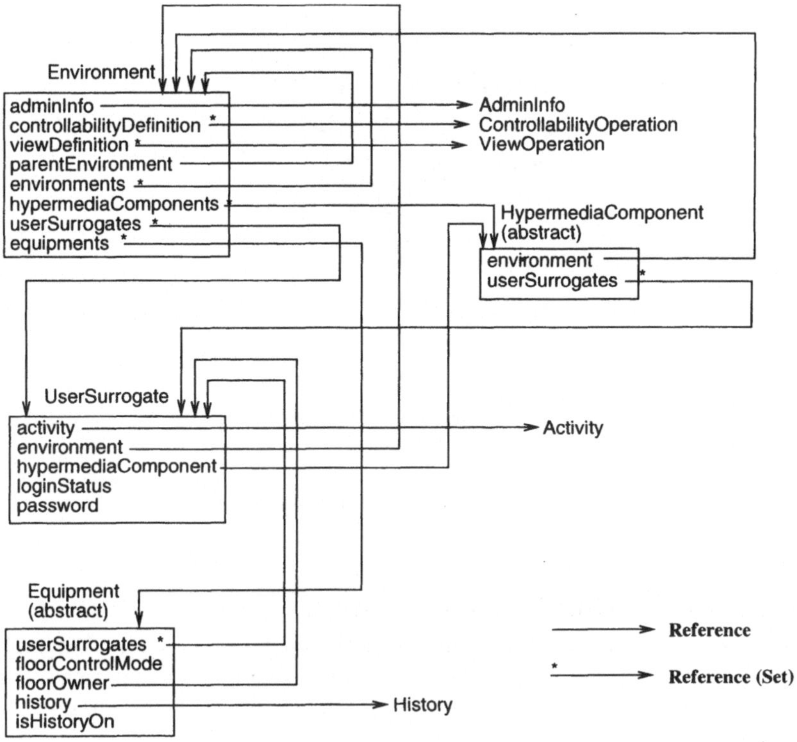

Figure 1: Main attributes of and relationships among the four fundamental objects

- *Environment objects* are containers of other fundamental objects. A user has access to objects in the same environment where his/her user surrogate object resides. An environment object can have child environments with inheritance relationship. Basically, definitions of a parent environment and objects contained in it are inherited by child environments. Therefore, users in an environment have access to objects contained not only in the environment but also in ancestor environments of it. Environment objects provide an environment model that plays the most important role in VIEW Media.

Figure 1 represents main attributes of and relationships among these four fundamental objects.

3.2 The Customization Mechanism of VIEW Media

VIEW Media provides the customization mechanism for hypermedia component objects. The mechanism is based on the Object Deputy Model[34].

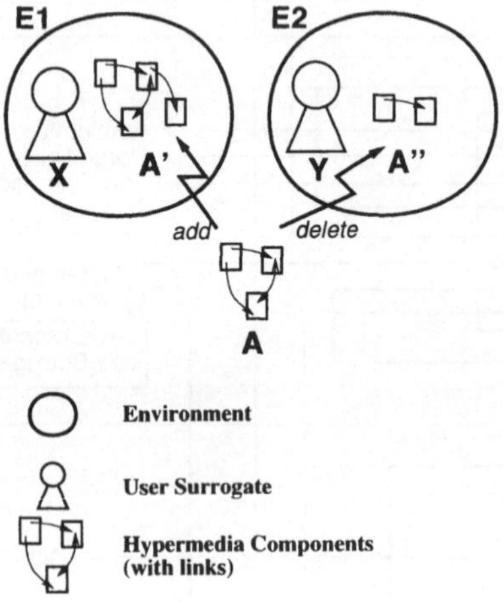

Figure 2: Customization mechanism

Environment objects have the most important role in the mechanism. An environment object has an attribute called *viewDefinition* as shown in Figure 1. The *viewDefinition* is an ordered set of customization operations to hypermedia component objects. When a user intend to refer to some hypermedia components, the environment object to which the user belongs behaves as follows:

1. fetching the required hypermedia components.
2. making the copies of them.
3. applying the set of customization operations stored in the *viewDefinition* to the copies.
4. providing the customized copies of hypermedia components for the user.

Thus, the customization does not affect the source hypermedia components and users always receive the customized copies of hypermedia components (Figure 2). It guarantees the independence of customizations in each environment. For example, if a user modifies some hypermedia component by customization operations, the modification only affects the users who belong to the same environment or its child environments.

3.3 WYSIWIS

Most groupware systems introduce WYSIWIS or location relaxed-WYSIWIS principles so that all users can see the same display contents. It enables users to share the same perception in contents of cooperative work and helps users to promote work smoothly. However, the principle does not suit all condition of cooperative work. For example, when a user wants to show only a part of his/her own document, because of security reason to others, or when a teacher wants to use a textbook with annotations while students use a textbook without them, the principle is too restrictive. In these cases, it is required that users can modify part of documents for flexible cooperative work. VIEW Media provides this flexibility and realizes cooperative work under the non-WYSIWIS condition.

3.4 Awareness

Providing awareness information becomes more important when users collaborate under non-WYSIWIS condition than WYSIWIS or relaxed-WYSIWIS condition. The degree of WYSIWIS, flexibility of collaboration, and needs of explicit awareness information are mutually related. The strict WYSIWIS view of a system enables users to assume that all other users have the same recognition and needs less explicit awareness information than relaxed-WYSIWIS or non-WYSIWIS, though flexibility of it is limited. The relaxed-WYSIWIS view permits users to see the different part of the same documents and relieves the limitation of the strict WYSIWIS view and provides flexibility of collaboration to some extent. It requires more awareness information than strict WYSIWIS. For example, an awareness support tool providing information that shows what part of documents each user browses, like shared pointers, will be required. The non-WYSIWIS view, having much potential of flexible collaboration, does not guarantee the recognition of users to be the same at any time, so that more explicit awareness information is needed than other two conditions.

3.4.1 Handling Awareness Information

A system supporting non-WYSIWIS view should have an explicit model to manage awareness information. VIEW Media introduces the notions of *controllability* and *observability* as generalization of awareness[36]. There are several reasons of modeling awareness information.

Controlling awareness information. Awareness information should be controlled by a system to provide precise awareness information in some cases. The "control" mainly means selecting or reducing the information for some reasons: (1) In general, users have different positions or responsibilities under non-WYSIWIS condition because of the flexibility of it, as opposed to users under WYSIWIS or relaxed-WYSIWIS condition who tend to have the same position or responsibility because of limitation of flexibility. It is required controlling awareness information for ensuring security or privacy of users in such a situation. (2) The number of users also will be a reason of controlling awareness information. When the number of users becomes large, providing awareness information of all users is not realistic, because the information may occupy the greater part of the screen. It is required filtering the information not much related with a user who receives the information.

To control awareness information, a system must handle the contents or semantics of the awareness information and the information about relationship among users. Thus, awareness information model that enables handling contents or semantics of the information and represents relationship among users is required.

Processing awareness information. Not only controlling but processing awareness information will be required by a system supporting the non-WYSIWIS principle. We propose grouping, taking statistics, and abstraction as typical processing methods. To realize such processing, the representation of awareness information should be easy to deal with. *Grouping* is used as a method of generating awareness information of groups. For example, when some groups work concurrently, users may want awareness information that indicates a status of each group roughly estimated instead of the information about each user. The system summarizes each attribute of awareness information about each user and shows the summarized information in a precise manner. Ordinarily, awareness information provides *current* statuses of users. In contrast to this, *statistical awareness information* summarizes a sequence of awareness information of objects (users or groups) from the past to now. The system needs to have a kind of history buffer for storing awareness information. *Abstraction* is also a method of ensuring security or privacy of users. This method extracts only necessary information from source awareness information and shows the extracted information in an abstract way. For example, a system extracts an activity of a user from video and audio data and shows the activity as a kind of symbolic characters representing the user. It controls the degree of exposure of awareness information so that undesirable information will be suppressed.

The principal usage of awareness information will be classified into two types: identifying characteristics of individual users and comparing characteristics among users. To support the latter usage, it is required to normalize awareness information for comparison. It will be achieved by applying methods described above.

3.4.2 Examples

We developed several tools providing awareness information based on the notion discussed above. Selected example tools are presented here.

Shared Pointer. Figure 3 shows the screen of a prototype system of VIEW Classroom, a distance learning system which have been developed in our laboratory[22]. Shared pointers are used in the window on the upper left side of the screen. The shared pointer can indicate explicit awareness information, such as a role or a name of a user with a text string, his/her activity with a color of the pointer. Each user has different pointer sharing status. Figure 3 represents a screen of a student (Iwamoto). The status of pointer sharing is shown in *Pointer Sharing Graph* window, located in the lower right side of the screen. It represents that "Students" can share pointers

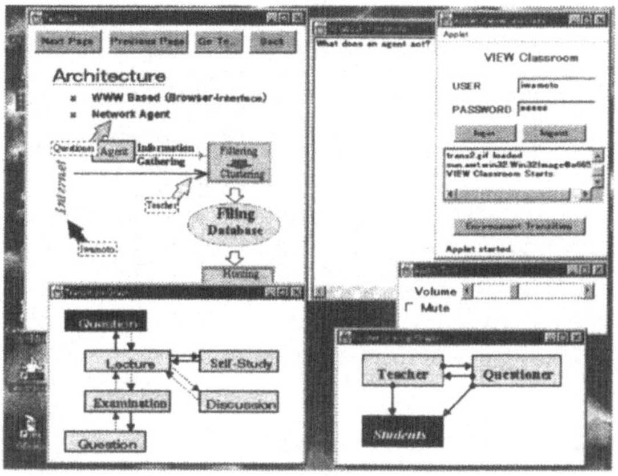

Figure 3: Shared pointers with explicit awareness information

of "Teacher" and "Questioner", however, pointers of "Students" are not shared with any other users. The status of pointer sharing is dynamically changed according to a situation of collaboration, indicated in *Environment Transition Graph* window located in the next side of the *Pointer Sharing Graph* window.

Virtual Seats. Virtual seats are also the feature of VIEW Classroom providing awareness information used by teachers. When the number of students attending a distant classroom session becomes large, a teacher will be unable to grasp statuses of all students and hesitate whom s/he should choose for questions. A teacher can issue a query that selects students to a system. The condition of the query includes a record of a student, the number of times a student speaks, and so forth. The system generates a window of virtual seats composed of selected students by the query. The teacher can use the virtual seats as representatives of students. For example, when 500 students exist in a lecture, a teacher can select 50 students satisfying some conditions. Each seat of virtual seats indicates a status of a student. Virtual seats also provide *abstract audio awareness information.* They transmit activity of students. For example, when many students are annotating a textbook, the sound of writing with a pencil will be generated. When many chat channels exist among students, the prepared sound of a chat will be played. These are examples of abstract awareness functions: they convert medium of awareness and abstract necessary information from source information. The teacher can reconstruct virtual seats by issuing a new query according to the change of a situation during a lecture.

Channel Checker. Figure 4 shows a conference support system with *Channel Checker*[3)], which provides awareness information. Figure 4 is a screen of a user in the second phase evaluation group. In the document browser, the contents written

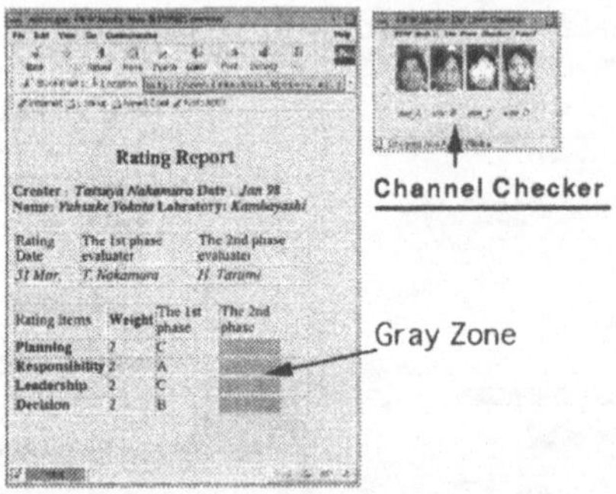

Figure 4: Information of access rights to shared documents

in the first phase evaluation are always seen by the user. The contents written in the second phase evaluation are masked with gray rectangles. When the user want to see the contents, s/he has to move his/her pointer into the rectangles, then the contents will be appear. This function is closely related with the functions of *Channel Checker*. The contents written in the last phase evaluation are not shown. Thus, the user does not know the existence of the contents.

When the user moves his/her pointer into the gray rectangles, *Channel Checker* shows who can see the contents of the part and the channel for voice conversation will be changed so that the user can speak only users who can see the contents. Therefore, when the user speaks about the contents masked with the rectangles, the contents of the conversation are not transmitted to users who cannot see the part and security or privacy will be ensured.

Environment and Awareness. Use of VIEW Media to realize a debate system is discussed in [37], where environment information is displayed for awareness purpose.

3.5 Implementation

VIEW Media has client-server architecture, as it is shown in Figure 5. All fundamental component objects are managed on the server.

VIEW Media is written in Java[27] (JDK1.1.5) with HORB[30] distributed object environment. Figure 6 shows the screen shot of VIEW Media client. Resource Manage Server runs on the host running WWW server as a HORB daemon object. Clients are realized as Java applets, thus users can use VIEW Media on a WWW browser without any installation procedure. VIEW Media clients provide these user interfaces as individual tools as follows: *Hypermedia Browser* displays hypermedia documents of VIEW Media. Users can navigate hypermedia by following links, and add or delete comments and links by making anchors. *Text Chat* supports conversation among users using texts. *Awareness Window* displays the structure of

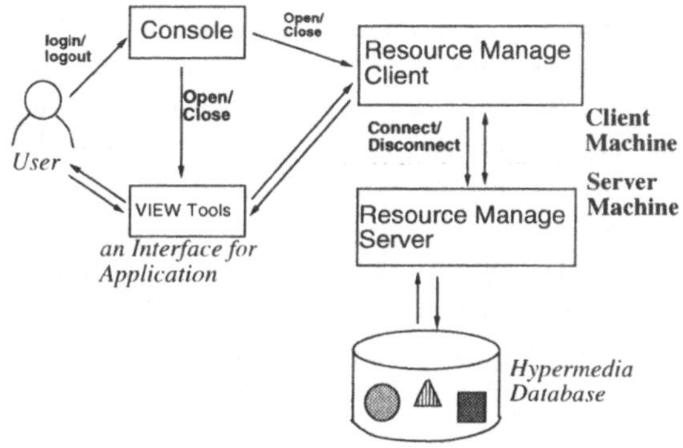

Figure 5: System Architecture

cooperative workspace which consists of environment objects, the present statuses of users, and so forth.

System developers can extend these tools or add new tools for purposes of development of various groupware.

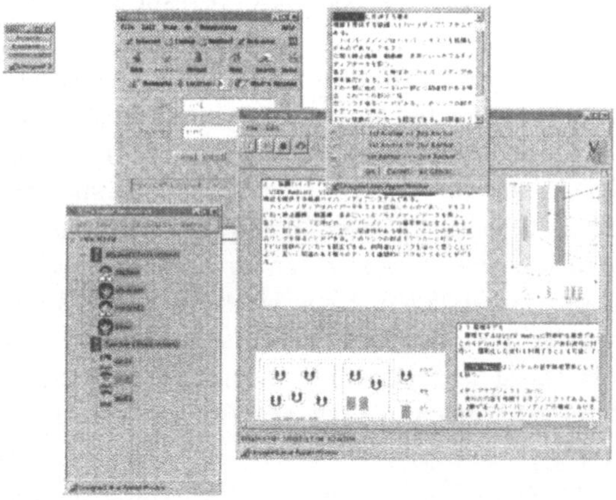

Figure 6: Screen Shot of VIEW Media

4 Related Work

BSCW (Basic Support for Cooperative Work) at GMD-FIT[7] is a web-based groupware system providing shared workspaces for users. Each shared workspace corresponds to each work group. Users can belong to one or more work groups, namely, workspaces. To join a workspace, users have to pass password authentication. Basically, a workspace is a repository of various types of files. In BSCW, users in the same workspace can have different access levels. This mechanism is realized by providing a per-object access control model. A notion of a *bag* supports migration of data among various workspaces. Bag is a personal repository and users always have access to their own bags to copy or move data among workspaces. Bag is regarded as a simple function to support personal work in BSCW.

TeamWave is a typical groupware system based on TeamRooms[8)] using a room model[9)]. The room model of TeamWave provides persistent *places* for collaboration as opposed to the temporal meetings provided by the meeting-centered models of many conventional real-time groupware systems. The notion of place provides features of physical rooms of the real world. Physical team rooms have various characteristics suited to collaboration. The room model simulates the characteristics on the networked computing environment. Rooms organize persistent workspaces for users containing documents and various tools. Rooms and their contents are fully persistent. Thus the model is able to support both real-time and asynchronous cooperative work. Multiple rooms are able to exist in a system and each room represents one kind of work. Users can move to other rooms easily to change their work. TeamWave integrates individual and group work and does not distinguish individual and group workspace. Rooms are able to be used for both individual and group work. Thus a transition between both types of work corresponds to a movement between rooms.

CBE (Collaboratory Builder's Environment)[10, 11)], also based on the room model, is a groupware system implemented in the Java language. Users of CBE can utilize the system through WWW browsers. CBE supports both personal work and group work simultaneously. Users can switch their work style smoothly. The basic unit of workspace in CBE is a *room*. The main purpose of rooms of CBE is providing workspace for cooperation and storage of data. Multiple rooms in CBE represent multiple work going on concurrently. Each workspace in CBE is able to define the roles of users. The roles are classified into four types: *Administrators*, *Members*, *Observers*, and *Restricted*. The notion of roles in CBE only represents the authority of users in each room. We think the roles of CBE are less expressive than those of the environment model.

The WORLDS project at DSTC (Distributed Systems Technology Centre) in Australia have developed Orbit, which is a system based on the *locale* framework[12)]. Locales, defined as places for groups working in various social worlds, are able to be regarded as a unit of workspace in Orbit. One social world corresponds to one locale and users can belong to one or more locales. A locale is a repository of data or tools for cooperative work. A user who belongs to some locales has an *individual view* that consists of tools and data selected from those locales. Thus individual views generally differ from each other even if users belong to exactly the same locales. This mechanism provides users who belong to one or more social worlds with an integrated view.

Suite[13] provides an access control model based on access matrices, it also supports dynamic roles. The access matrix implements an inheritance mechanism reducing the specification costs. The mechanism has conflict resolution rules in the subject, object, and access right dimensions. The access control model can describe negative rights in addition to traditional positive rights. Negative rights also reduce the specification costs of supporting dynamic roles.

Intermezzo[14, 15], a toolkit for groupware systems, provides an access control model based on the notion of *roles* and *policies*. It also supports static and dynamic roles. Static roles are represented as lists of explicitly defined users. Dynamic roles, introduced for supporting dynamics and flexibility of collaboration, are represented as predicate functions using statuses or attributes of users as parameters. It realizes defining lists of users implicitly. Dynamic roles can determine membership in the role at runtime. A policy is a set of access control rights. Roles and policies are combined by many-to-many mappings. It provides sharing policies among roles and multiple policies associated with one role. The environment model of VIEW Media also supports static and dynamic roles, which integrates the notion of *roles* and *places*.

5 Conclusion

There are a lot of functions of database management systems applicable to practical CSCW systems. Sharing data among users and applications is a basic function of CSCW systems. Persistency of data realizes long-time collaboration and reuse of data. The capability of customizing data provides flexibility for users. CSCW systems require active capability to develop various event-driven services.

VIEW Media is a framework for CSCW systems developed by our laboratory. It adopts database technology as a foundation of its architecture. VIEW Media introduces functions of databases mentioned above and realizes flexible cooperative work. Developers can build applications for specific purposes on the system, such as a distance education system, a distributed conference system, and so forth.

Bibliography

1) H. Takada and Y. Kambayashi. An object-oriented office space description model and a office view management mechanism for distributed office environment. *Proc. 4th Int. Conf. Foundations of Data Organization and Algorithms*, October 1993, 362–377

2) M. El-Sharkawi and Y. Kambayashi. Object migration mechanisms to support updates in object-oriented databases. *Proc. of PARBASE-90: the Int. Conf. on Databases, Parallel Architectures, and Their Applications*, March 1990, 378–387, Also in *DATABASES: Theory, Design and Applications*, (Rishe, Novathe, Tal (Eds.)), IEEE Computer Science Press, 1991, 73–92.

3) T. Nakamura, Y. Yokota, S. Konomi, H. Tarumi, and Y. Kambayashi. A conference user interface supporting different access rights to shared hypermedia. *Proc. of the Asia Pacific*

Conference on Computer Human Interaction (APCHI'98), Kanagawa, Japan, July 1998, 38–43.

4) H. Linaae, I. Wiedemann Lovseth, Y. Kambayashi. An authorization mechanism for decentralized dynamic security management. *Proc. of Third Euro-Japan Seminar on Information Modeling and Knowledge Base*, June 1993.

5) Y. Kambayashi, Q. Chen, and T. Kunishima. Coordination manager: A mechanism to support cooperative work on database systems. *Proc. 2nd Far-East Workshop on Future Database Systems*, April 1992, 176–183.

6) S. Konomi, Y. Yokota, K. Sakata, and Y. Kambayashi. Cooperative view mechanisms in distributed multiuser hypermedia environments. *Proc. 2nd IFCIS Int. Conf. on Cooperative Information Systems*, June 1997, 15–24.

7) R. Bentley, W. Appelt, Busbach. U., E. Hinrichs, D. Kerr, S. Sikkel, J. Trevor, and G. Woetzel. Basic support for cooperative work on the world wide web. *International Journal of Human-Computer Studies: Special issue on Innovative Applications of the World Wide Web*, 46(2):827–846, June 1997.

8) M. Roseman and S. Greenberg. Teamrooms: Network places for collaboration. *Proc. the ACM 1996 Conf. on Computer-Supported Cooperative Work*, 1996, 325–333.

9) S. Greenberg and M. Roseman. Using a room metaphor to ease transitions in groupware. Research report 98/611/02, Department of Computer Science, University of Calgary, January 1998.

10) J. H. Lee and et al. Supporting multi-user, multi-applet workspace in cbe. In *Proc. the ACM 1996 Conf. on Computer-Supported Cooperative Work*, 1996, 344–353.

11) A. Prakash, H. S. Shim, and J. H. Lee. Data management issues and tradeoffs in CSCW systems. *IEEE Transactions on Knowledge and Data Engineering*, Vol. 11, No. 1, January/February 1999, 213–227.

12) T. Mansfield, S. Kaplan, G. Fitzpatrick, T. Phelps, M. Fitzpatrick, R. Taylor. Evolving Orbit: A progress report on building locales. *Proc. the Int. ACM SIGGROUP Conf. on Supporting Group Work : The Integration Challenge (Group97)*, November 1997, 241–250.

13) H.-H. Shen and P. Dewan. Access control for collaborative environments. *Proc. of ACM 1992 Conf. on Computer-Supported Cooperative Work*, 1992, 51–58.

14) W. K. Edwards. Session management for collaborative applications. *Proc. the ACM 1994 Conf. on Computer-Supported Cooperative Work*, 1994, 323–330.

15) W. K. Edwards. Policies and roles in collaborative appplications. *Proc. the ACM 1996 Conf. on Computer-Supported Cooperative Work*, 1996, 11–20.

16) G. Mark and et al. Hypermedia structures and the division of labor in meeting room collaboration. *Proc. the ACM 1996 Conf. on Computer-Supported Cooperative Work*, 1996, 170–179.

17) M. Sohlenkamp and G. Chwelos. Integrating communication, cooperation and awareness: The DIVA virtual office environment. *Proc. the ACM 1994 Conf. on Computer-Supported Cooperative Work*, 1994, 331–344.

18) G. Fitzpatrick, S. Kaplan, and T. Mansfield. Physical spaces, virtual place and social worlds: A study of work in the virtual. *Proc. the ACM 1996 Conf. on Computer-Supported Cooperative Work*, 1996, 334–343.

19) P. J. Spellman, J. N. Mosier, L. M. Deus, and J. A. Carlson. Collaborative virtual workspace. *Proc. the Int. ACM SIGGROUP Conf. on Supporting Group Work : The Integration Challenge (Group97)*, November 1997, 197–203.

20) R. B. Smith, R. Hixon, and B. Horan. Supporting flexible roles in a shared space. *Proc. the ACM 1998 Conf. on Computer-Supported Cooperative Work*, 1998, 197–206.

21) H. Iwamoto, C. Ito, and Y. Kambayashi. Design and implementation of action history view mechanisms for hypermedia systems. In *Proc. 21nd Annual Int. Computer Software and Applications Conference*, pages 412–420, Vienna, Austria, Aug. 1998. IEEE Computer Society Press.

22) Iwamoto, H. and Kambayashi, Y.: Dynamic control mechanisms for pointer share. Proc. the 13th Int. Conf. on Information Networking (1998) 10A-2.1–6

23) M. Stefik, D. Bobrow, G. Foster, S. Lanning, and D. Tatar. Wysiwis revised: Early experiences with multiuser interfaces. *Trans. Office Information Systems*, 5(2):147–167, 1987.

24) H. Abdel-Wahab, B. Kvande, and S. Nanjangud. Using java for multimedia collaborative applications. In *Proc. PROMS'96: 3rd Int. Workshop On Protocols for Multimedia Systems*, pages 49–62, Madrid, Oct. 1996.

25) J. B. Begole, C. A. Struble, C. A. Shaffer, and R. B. Smith. Transparent sharing of java applets: A replicated approach. In *Proc. the 1997 Symposium on User Interface Software and Technology (UIST'97)*, pages 55–64, NY, 1997. ACM Press.

26) P. Dourish and V. Bellotti. Awareness and coordination in shared workspaces. In J. Turner and R. Kraut, editors, *Proc. 4th Int. Conf. on Computer-Supported Cooperative Work*, pages 107–114, Toronto, Canada, Nov. 1992. New York: SIGCHI/SIGOIS ACM.

27) J. Gosling, B. Joy, and G. Steele. *The Java Language Specification*. Sunsoft Java Series. Addison-Wesley Developers Press, 1996.

28) F. Halasz and M. Schwartz. The dexter hypertext reference model. *Communications of the ACM*, 37(2):30–39, Feb. 1994.

29) R. Hazemi and L. Macaulay. Requirements for graphical user interface development environments for groupware. *Interacting with Computers*, 8(1):69–88, 1996.

30) S. Hirano. Horb: Distributed execution of java programs. In *Worldwide Computing and Its Applications(Springer Lecture Notes in Computer Science 1274)*, pages 29–42, 1997.

31) K. Katayama, O. Kagawa, Y. Kamiya, H. Tsushima, T. Yoshihiro, and Y. Kambayashi. Use of action history views for indexing continuous media objects. In S. Nishio and F. Kishino, editors, *Proc. 1st Int. Conf. on Advanced Multimedia Content Processing*, pages 349–360, Osaka, Japan, Nov. 1998.

32) J. H. Lee and et al. Supporting multi-user, multi-applet workspace in cbe. In *Proc. the ACM 1996 Conf. on Computer Supported Cooperative Work*, pages 344–353.

33) G. Mark, J. M. Haake, and N. A. Streitz. The use of hypermedia in group problem solving: An evaluation of the dolphin electronic meeting room environment. In *Proc. the 4th European Conference on Computer-Supported Cooperative Work*.

34) Z. Peng and Y. Kambayashi. Deputy mechanisms for object-oriented databases. In *Proc. IEEE 11th Int. Conf. Data Engineering*, Mar. 1995.

35) Y. Yokota, K. Sugiyama, H. Tarumi, and Y. Kambayashi. Evaluation of non-wysiwis functions of view media. In *Proc. 2nd Int. Symposium on Cooperative Database Systems for Advanced Applications (CODAS'99) (to be appeared)*, Mar. 1999.

36) Yokota, Y. and Kambayashi, Y.: Customization support environments of active hypermedia systems for cooperative work. Advanced Database Systems for Integration of Media and User Environments '98, World Scientific (1998) 21–26

37) Yokota, Y., Sugiyama, K., Tarumi, H. and Kambayashi, Y.: Evaluation of Non-WYSIWIS Functions of VIEW Media, Proc. 2nd Int. Symposium on Cooperative Database Systems for Advanced Applications (CODAS'99), Springer (1999) 88–99

4

Broadcasting and Databases

Katsumi Tanaka

Graduate School of Science & Technology, Kobe University

Kazutoshi Sumiya

Research Center for Urban Safety & Security, Kobe University

Akiyo Nadamoto

Graduate School of Science & Technology, Kobe University

Qiang Ma

Graduate School of Science & Technology, Kobe University

ABSTRACT

Recently, much attention has been focused on digital TV broadcasting and broadcast-type information delivery systems. Notable features of the broadcast-type information delivery systems are that the same information is transmitted to anonymous users, and that users receive those information in a passive manner. The delivered contents is usually composed of multiple components, whose update timings may vary. This paper describes our recent research focussing on the above substantial aspects of broadcasting-type information systems. First, we will describe a version management mechanism to maintain the *temporal consistency* of delivered data in data broadcasting systems. Secondly, we will describe a way to view Web contents in a *passive manner* like TV-program contents. Finally, we will show our new information filtering mechanism for data broadcasting systems considering the *time-series features* of data, especially, the *freshness* and the *popularity* of data.

1 Introduction

Recently, much attention is focused on data broadcasting systems via the internet or digital broadcasting because of their potential and convenience. The concept of these systems is generally based on the *push-type information delivery* technology. The push-type information delivery does not require that users behave in an active manner in order to access information resources because the required and/or updated information is transmitted continuously and automatically. Users select their favorite channels provided by a service server in advance, and then the system will transmit information for each user at regular intervals. Recently, several push technologies, such as Pointcast 1.0[1)] and Castanet 2.0[2)], have been developed. These systems are not only applied to transmit text data but also hypertext and hypermedia data. In these data broadcasting systems, where data at the server side is often updated, one of the important problems is how to maintain the *temporal consistency* of the transmitted data because not all the transmitted data is received by the clients since client equipments are not always powered on. In this paper, first, we will describe our new version control mechanism by which the temporal consistency is kept at both the server and its clients[3, 4, 5)].

Rapid progress in the Internet, especially Web technology, brought a powerful information environment. Now, Web is regarded as the largest information resource. The Web browsing operation is easy for computer users, but it is still difficult for non computer users and ordinary TV audiences. Current Web browsers enforce users to have knowledge about computer operations. In order to acquire necessary information, users must *read and scroll* a Web page and *click and navigate* its hyperlinks. We call such a user browsing interface a *read and click interface.* On the other hand, TV audiences usually *watch and listen* to the TV programs in a passive manner, and acquire necessary information. Our idea is to introduce the *passiveness* into Web data viewing environment[6, 7)]. We propose a new type of Web viewing to obtain information from the Web. We call this interface a *watch and listen* interface. The proposed watch and listen interface presents Web pages like ordinary TV programs, where the content is presented by CG character's speech and images appearing in Web pages.

Broadcasting-type information dissemination systems on the Internet are becoming increasingly popular due to advances in the area of Web technology and information delivery. One of the notable features of push-based, multiple-channel-based information dissemination systems is to send information to users in a form of time-series articles. Conventional information filtering method does not consider well the *worth* of an article from the standpoint of the *time-series feature.* Finally, in this paper, we will describe our new information filtering mechanism[9, 10)] which considers the worth of an article compared with past delivered articles, that is, their time-series features (freshness, popularity, and urgency).

2 A Version Control Mechanism for Data Braodcasting Systems

2.1 Problems

In data broadcasting systems, contents are often updated and delivered periodically. All versions of contents are stored on the server whenever they are changed, and all versions received by clients are assumed to be stored in the clients. The clients are not neccesarily able to receive all the versions since the clients are not always

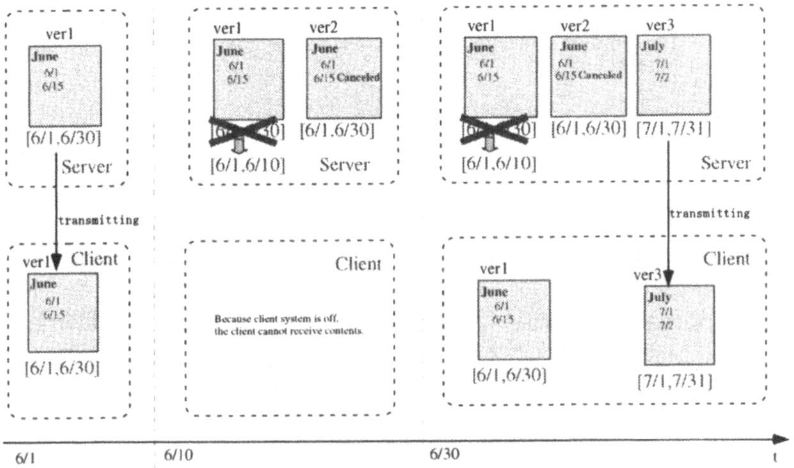

Figure 1: The problem of broadcast temporal information

powered on. Thus, some versions may not be able to be received by some clients. This leads to the possibility of a temporal inconsistency among stored contents at client sides. That is to say, the problem of maintaining the temporal consistency of contents in data broadcasting systems is caused by the fact that some clients cannot always receive all the versions.

Figure 1 shows an example, in which some valid times of versions stored at a client become inconsistent because the client could not receive all the versions. In this example, each rectangle and each time interval (e.g. [6/1,6/30]) associated with it denote a data unit and the data unit's valid time, respectively. Here, the data unit [The schedule on June] was first created and transmitted to a client on June 1st. The valid time of this data unit was originally assumed to be [6/1,6/30]. At the server side, assume that a certain event (on June 15th) in the data unit was cancelled on June 10th. Since this data unit is updated, the valid time of the previous version (say version 1) of the data unit is changed to [6/1,6/10]. Assume that the client equipment was, however, not powered on at the time, and that the client could receive neither the updated data unit (version 2) nor the document (version 1) whose valid time was changed. Then, if the data unit [The schedule on July] whose version number is 3 is created and transmitted on June 30, the valid time of the data unit (version 1) stored at the client becomes to contain wrong information and leads to cause a temporal inconsistency.

2.2 Version Trees for Updates of Contents and Valid Times

Here, we propose a version control mechanism[5)], in which for each data unit,

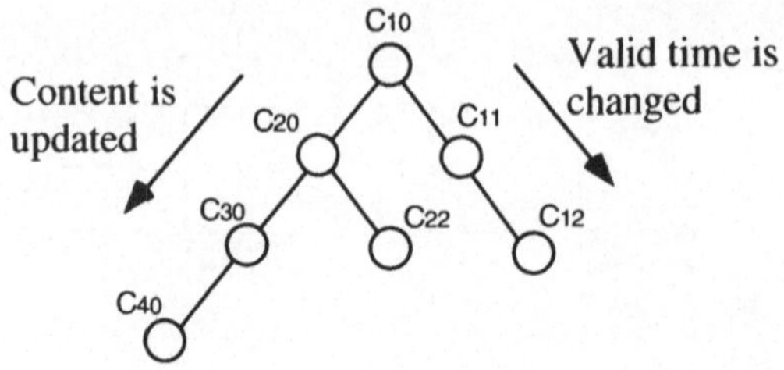

Figure 2: A Version Tree Representing Updates of Content and Valid Time

both of updates of it's content and valid time are represented by a binary tree at the server. In this mechanism, we limit our version model to the linear version model[1]. Figure 2 shows an example version tree for a certain data unit. When the data unit's content is updated and a new version is produced, the new version is represented as a left child node. When the valid time of the data unit is updated and a new version is produced with the same content and the different valid time, the new version is represented as a right child node. The server transmits all leaf nodes of this version tree, and the client can store only the versions whose valid times are correct. In this mechanism, we assume that whenever the content is updated and the new version is created, the valid time of the original data unit version is shortened.

A version tree for a data unit is defined as follows:

$$\begin{aligned} T &= (N, E) \\ N &= \{c_1, c_2, \cdots, c_n\} \\ E &\subseteq \{(c, c')|c \in N, c' \in N\} \\ c &= (id, version, tr, [v_1, v_2]), c \in N \end{aligned}$$

Here, N denotes the set of versions (nodes) and E denotes the set of edges representing version derivations. Each left edge denotes an update of content, and each right edge denotes an update of valid time. c is a node of the version tree T, which denotes a version of the data unit. id is the identifier, $version$ is the version number, tr denotes the transaction time when the version is registered at the server. $[v_1, v_2]$ denotes the valid time of a version of the data unit.

Whenever the content (and possibly its valid time) of a data unit is updated, a new version is created and added to the correponding version tree as a new left child node of the updated node. Whenever only the valid time of a some version of a data unit is changed, a new version is also created and added to the version tree as a new right child node. Figure 3 shows two cases of data unit updates. Figure 3(a) shows the case when both the content and its valid time are updated. Assume that

[1]The linear version model means that version derivation relationships are linear. That is, we assume that each version generation for content update and valid time update is done in a linear manner.

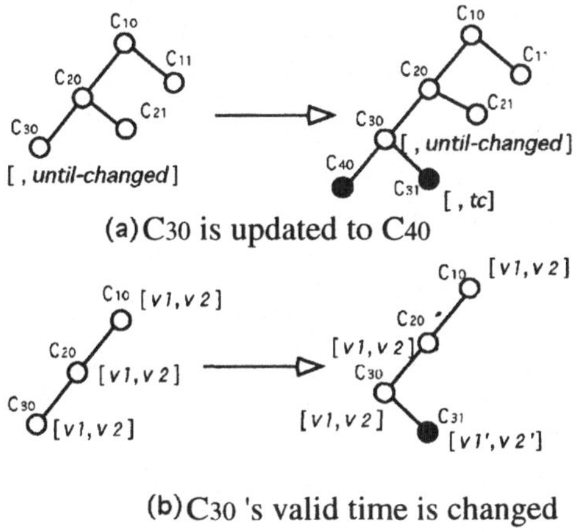

Figure 3: Updating the unit at time t_c

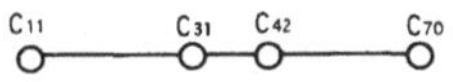

Figure 4: A Version List at a Client

before the update, the newest version was C_{30} whose expiration time is assumed to be *until – changed*. After the update is done, the node C_{40} (which is changed from C_{30}) and C_{31} (whose expiration time is t_c) are added to the version tree, because C_{30} is now invalid. Figure 3(b) shows the case when changing only the valid time of the node C_{30}. A new right node C_{31} is added to represent the change of the valid time.

Whenver the server receives a request for transmission of a data unit by a client, the server transmits all the leaf nodes of the corresponding version tree to the client. The client receives all the leaf nodes, and then creates a *version list* in order to control versions. A version list represents a list of versions received by the client. Formally, a version list is defined as follows:

$$l = [c_1, c_2, \ldots, c_n]$$
$$c = (id, version, tr, [v_1, v_2], ar)$$

Here, c is a version of the data unit. *id*, *version*, *tr*, and valid time are the same as the server's ones. *ar* means the arrival time of the node. It should be noted that there may be some nodes that cannot be received while a client system is off.

At the client, when a new version is transmitted from the server, the following process is carried out.

- Creation of a new version list
  If the version list fitting the received node is not found, a new version list is created and the received node is added to it.

- Addition of a received node to a version list
  If the received node's *major version number*[2] i is the largest, the node is added as the last part of the version list.

- Insertion of a received node to a version list
  If the major version number i of the received node is not equal to any node's version number of the version list, the node is inserted.

- Exchanging old versions by new ones.
  If the received node's major version number i is equal to one of the node's version number and the received node's *minor version number* is larger than the node, the existing nodes are replaced by the new nodes. The old version is deleted because its valid time is wrong.

Figure 5 shows the process at the client.

1. The node c_{10} is created at the server. The client receives the node, and makes a new version list.

2. The node c_{10} is updated and c_{20} is created, c_{11} is added because the valid time of c_{10} is changed. At the time, the client could not receive them because the client system is off.

3. The node c_{20} is updated and c_{30} is created at the server. The client system is powered on and requests the server for transmission. The server transmits all the leaf nodes c_{30} and c_{11} to the client. The client receives these nodes. The node c_{30} is added to the version list, and c_{10} is exchanged by c_{11}.

4. The node c_{30} is updated and c_{40} is created at the server. c_{21} and c_{31} are created and added to the version tree, because the valid times of c_{20} and c_{30} are changed. The client receives the leaf nodes c_{40}, c_{21}, c_{31}. The node c_{21} is inserted, c_{31} is exchanged for c_{30}, and c_{40} is added to the version list of the client.

When a client system is powered off, the client cannot receive nodes. The nodes whose valid times are changed at the server are transmitted to the client when the client system is powered on, because the server transmits all the leaf nodes of the version tree. If necessary, the client exchanges previously delivered nodes by newly

[2]Here, we assume each version number consists of two components called *major* and *minor* version numbers. The major version number denotes the version concerned with updates of content. The minor version number denotes the version concerned with updates of valid times. In our examples, two digit numbers are used as version numbers, where the leftmost digit denotes the major number and the rightmost number denotes the minor number.

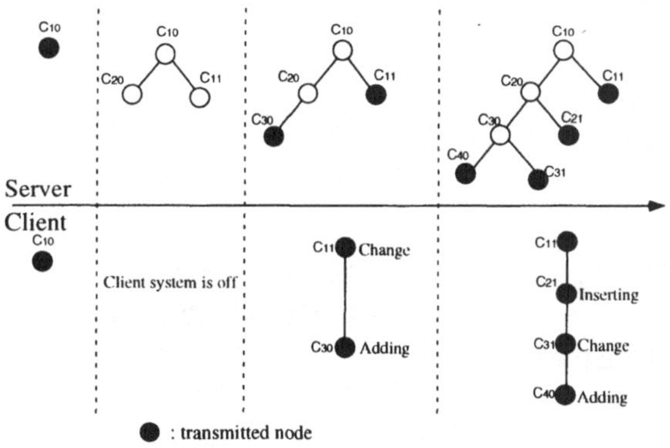

Figure 5: Process at the client

delivered nodes, and stores only nodes whose valid time are correct. Therefore, although there are some nodes which cannot be received, the client can keep the temporal consistency of stored nodes.

3 Passive Viewing of Web Content

In this section, we describe a way to browse Web pages in a more passive style[6, 7]. As shown in Figure 6, the proposed *watch and listen interface* is our solution for this objective. Once a user specifies a URL, the Web page corresponding to the URL is shown like as a TV-program, which the user can watch and listen to it. Figure 7 shows an example screen image for the watch and listen interface. The left part of Figure 7 is an ordinary Web page containing texts, images and so on. The right part of Figure 7 is a screen image of our watch and listen interface. The interface presents the content of the corresponding Web page like a TV program, in which some animation characters speak the text in the original page like lines, and the images contained in the Web page are presented consecutively one by one, synchronizing with characters' speech and behaviors. In order to realize the *passive viewing of Web content*, we have been developing the following functions:

- **Automatic transformation of Web pages into TV-program-like content**
  By analyzing the tag structure and guessing synchronizable regions of a specified Web page, we have been developing a tool to automatically transform the Web page into a TV-like program (which is represented as a TVML[8)3] script).

[3]TVML (TV Making Language) is a language to describe the whole TV program that has been developed at NHK and the TVML consortium. TVML has several functions to make TV programs such as CG animation, speech synthesis, video replay control, and camera work control although it does not work in Web environment.

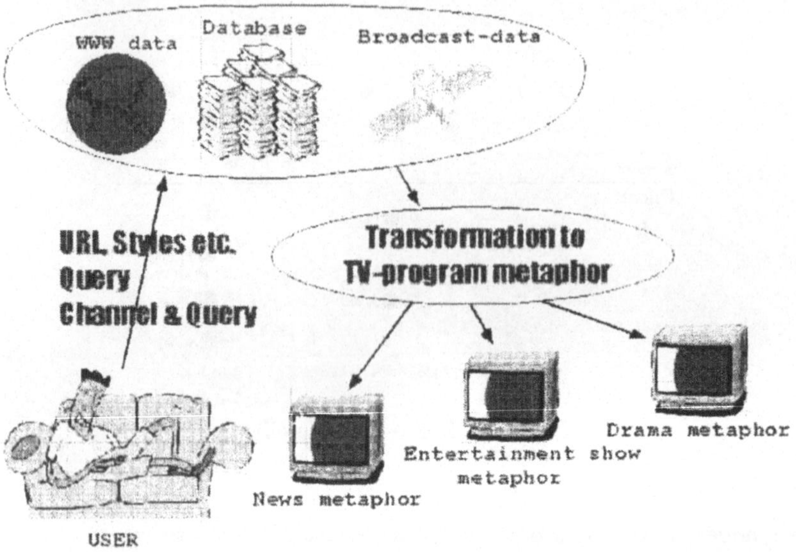

Figure 6: Passive Viewing of Data

- **Semi-automatic transformation**
  We have also been developing an authoring mechanism, by which authors of Web pages transform their Web pages into script-like contents. Partially, the above automatic transformation mechanism is used, but complementarily, authors explicitly specify the presentation styles, synchronization of images and texts, text to lines/behaviors conversions in a manual and explicit manner. The result of the semi-automatic transformation can be quickly reviewed as a TVML program. Also, the authoring result can be stored as a S-XML program (described below).

- **Preparing & using a markup language**
  This approach is to use a new markup language (tag extensions by XML), by which Web pages suitable for passive browsing can be created. For this purpose, we have been designing and implementing a language called Scripting-XML (abbreviated by S-XML).

3.1 Automatic Transformation for Passive Viewing

Our prototype system downloads a Web page by a user-specified URL into the system, and then, automatically transforms it into a series of TVML scripts. Then, the system automatically plays the generated series of scripts like a TV program by a user-specified TV-program style. Figure 7 shows both of a conventional Web browsing and our *watch and listen*-type viewing. The right part of Figure 7 shows

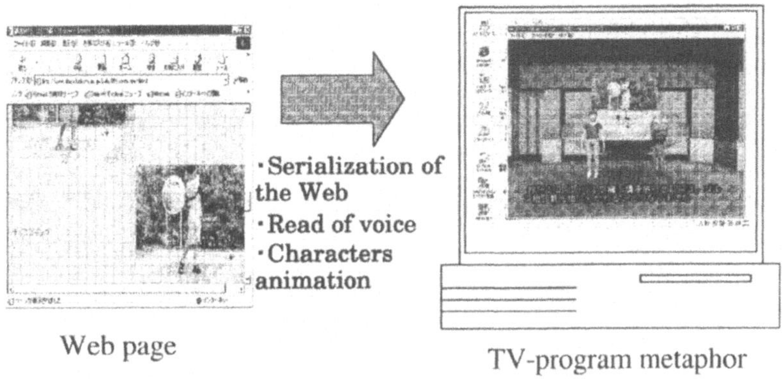

Figure 7: Transformation of a Web page into a TV-program-like content

the screen for the passive viewing, where the transformed Web page is played like a TV program. The original image appearing in the Web page is placed on the panel, and two character agents explain the image. The two characters also behave some performances (e.g. point out to the panel, walk around the panel etc.) during their speech. Basically, text portion in a Web page is used as a character's lines, and is spoken by the character. Images are basically placed on the panel appearing in the TV studio set. Since the Web pages have their own structures that are represented by HTML tags, our system automatically analyzes the structure, and transforms the content of a Web page into a TV-program-like content according to the user-specified TV-program style.

Our approach requires to transform the content of a Web page into a synchronized data streams by time code. When nothing considered, images and it's explanation text in the Web page are not automatically synchronized. This is because HTML itself does not have any tag concerned with synchronization of components. Therefore, it is necessary to guess which text synchronizes with an image. We explored an algorithm to discover possible synchronizable regions, in which a text describes an image. In finding such synchronizable regions, we mainly consider to find the *minimal tagged region*, which intuitively means a minimal portion of a Web page that is surrounded by a tag $< X >$ and $< /X >$ and that contains a specified image and some sentences. Usually, table tags of HTML are used to represent the synchronization between an image and its explanational text. Our algorithm to find minimal tagged regions can cope with the case of using those table tags.

3.2 Scripting-XML (SXML)

In order to create TV programs, we need to consider not only the content of the TV program, but also the presentation styles such as casting, camera movements, lighting, and characters' performances. Conventional markup languages such as HTML and XML mainly focused on creating structured documents that can be delivered in the Web environment. Therefore, they do not consider well the following issues:

- Line Assignment: Which portion of the content should be presented by what character?
- Synchronized Presentation: Which information should be presented simultaneously?
- Presentation Styles: Which style should be applied to represent the content?

Indeed, as a special type of structured documents, TV scripts or movie scripts have been traditionally treating the above issues. But, once those materials are represented as Web pages, they become sole textual data, and they have no effective browsing tool to express the above issues. In summary, the present HTML is not designed to describe TV scripts or movie scripts, but to describe ordinary documents.

Therefore, we have been developing a new markup language called Scripting-XML (abbreviated by S-XML) that is a collection of XML tags for describing scripts as Web pages. Scripts written in S-XML are automatically translated into TVML scripts, and can be played by the TVML player. In designing our S-XML, we separated the *content-related tags* from *style-related tags*. This is because we wish to allow the possibilities that scripts might be presented in more than one styles. Also, we wish the possibilities of the usage of those scripts as ordinary Web documents. By this separation, the content-related tags mainly specify which portion are used in the TV-like presentation. The style-related tags mainly specify how the specified content should be presented.

Figure 8 shows an example script written in S-XML. This script tries to introduce Web pages containing memorial photos with comments. The author specifies to present the content in a variety program metaphor. Two characters called BOB and MARY are used to present the content. The script mainly consists of four parts called `introduction, development, turnstory` and `conclusion`. When the part changes, for example, when the part moves from the introduction part to development part, the camera work and characters' movement try to express the change of the parts by atmosphere. The readers will find the attribute called `imgnum` and `refimgnum`, which appear in the tags `img` and `line`, respectively. The attribute `imgnum` denotes the image number, by which the attribute `refimgnum` of the `line` tag denotes that the designated image should be showed in a synchronized manner with the speech of the line.

4 Information Filtering of Time-series Articles by Freshness and Popularity

Broadcasting-type information dissemination systems on the Internet are becoming increasingly popular due to advances in the area of Web technology and information

```
<?xml version="1.0"?>
<!DOCTYPE tvprogram SYSTEM "S-xml.dtd">
<tvprogram>
<title>Tanaka's photo album about DBWS'99</title>
<metaphor>
<character model="BOB" name="BOB" voice="e_youngman"/>
<character model="MARY" name="MARY" voice="j_woman"/>
<variety>
<visiblechara refname="BOB"/>
<visiblechara refname="MARY"/>
<intro>
   <line> Hi! Dear, This is an photo album of the
   database workshop held in Okinawa which was
   taken by Professor Tanaka.
   </line>
</intro>
<development>
   <subject>
   Welcome party held in the second day.
   </subject>
   <img src="kiyoki.jpg" x="0.0" y="2.0" z="-3.0" imgnum="1"/>
   <line refimgnum="1">
   This person is Prof. Kiyoki, Keio University,
   who is currently the chairperson of IPSJ SIGDBS.
   </line>
   ......
</development>
<turnstory>
   <subject>
   Attractions at the dinners.
   </subject>
   <img src="buyou.jpg" x="0.0" y="2.0" z="-3.0" imgnum="6"/>
   <line refimgnum="6">
   This is the Okinawa dance show on the first day's dinner.
   </line>
   ......
</turnstory>
<conclusion>
   <line>
   Did you have good memories in Okinawa? OK.
   So, see you next time! Bye bye,..
   </line>
</conclusion>
</variety>
</metaphor>
</tvprogram>
```

Figure 8: A Script written in S-XML

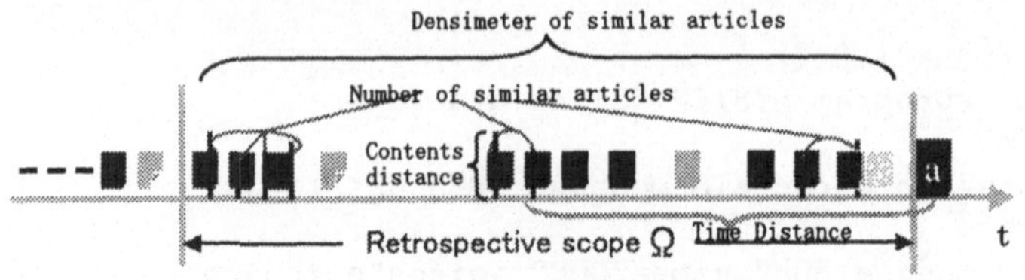

Figure 9: Freshness of Articles

delivery. One of the notable features of push-based, multiple-channel-based information dissemination systems is to send information to users in a form of time-series articles.

To find interesting information for users from the large quantity of data, information filtering techniques and search engines, which are mainly based on the keywords, have been very useful. However, since the keywords of incoming news articles are sometimes unknown, these typical methods may fail in acquiring the *fresh* (or *popular*) articles. The *freshness*, *popularity* and *urgency* are defined here as *time-series features* of news articles[9, 10)]. These features can be used to filter the time-series articles to acquire the fresh, popular and urgency news.

4.1 Freshness

The articles, which are quite different from previously selected articles, would be valuable. In other words, we can say that the articles have their freshness and uniqueness. Indeed in some cases, the articles may be *scoop* news.

As shown in Figure 9, the freshness of the article a can be estimated by

- the number of its similar articles in a restrospective scope, denoted by $fresh_{num}(a)$,
- the dissimilarity between a and the past articles in a retrospective scope, denoted by $fresh_{cd}(a)$,
- the densimeter of its similar articles in a retrospective scope, denoted by $fresh_{de}(a)$, and
- the time distance of a and its similar articles in a retrospective scope, denoted by $fresh_{td}(a, \omega)$.

The *integrated freshness* of an article a compared with articles in a retrospective scope Ω, denoted by $fresh_{\Omega}(a)$, is also defined as follows:

$$fresh_{\Omega}(a) = \alpha * fresh_{num}(a) \quad (4.1)$$
$$+\beta * fresh_{cd}(a, \omega) \quad (4.2)$$
$$+\gamma * fresh_{de}(a) \quad (4.3)$$
$$+\sigma * fresh_{td}(a, \omega) \quad (4.4)$$

where $\alpha, \beta, \gamma, \sigma$ are the weight values. ω is the set of articles similar to a that are in the scope Ω.

Let the m be the number of articles in ω and n be the number of articles in Ω. The above four types of *freshness* measurements are defined as follows.

Freshness based on the number of similar articles When there are few articles that are similar to a in Ω, we can say a is newer one and its freshness is considered to be high. So, we have

$$fresh_{num}(a) = \frac{1}{log_2(2+m)} \tag{4.5}$$

Freshness based on the content distance The *content distance* of article a and b can be defined as follows:

$$dis(a,b) = |v(a) - v(b)| \tag{4.6}$$

The content distance can represent how much new information has been added to a compared with b. Therefore, we can say, the content distance between a and its similar articles is becomes bigger, the freshness of a is considered to become higher. Thus, we have

$$fresh_{cd}(a,\omega) = \log(\frac{1}{m} \sum_{i=1,b_i \in \omega}^{m} dis(a,b_i)) \tag{4.7}$$

Freshness based on the densimeter of similar articles The *densimeter* d of the similar articles of a in Ω is computed as m/n. Here, d can be considered as the appearance probability of a in Ω. When d is small, a is rare one, and its freshness is considered to be high. So, we have

$$fresh_{de}(a) = log_2\frac{n}{m} \tag{4.8}$$

Freshness based on the time distance Assume that some articles in the past received archive are similar to article a and that the time distance between a and those similar articles is large. In this case, some fresh information is considered to occur, and so, the article a is considered to have a high freshness. So, we have

$$fresh_{td}(a,\omega) = \log(\frac{1}{m} \sum_{i=1,b_i \in \omega}^{m} (t(a) - t(b_i))) \tag{4.9}$$

where $t(a)$ is the broadcast time of article a.

4.2 Popularity

In order to select valuable one from the large quantity of news articles, the similarity and dissimilarity of the article compared with previously selected articles should

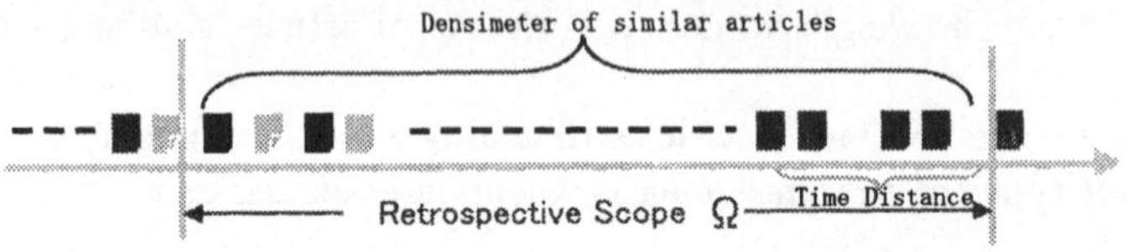

Figure 10: Popularity

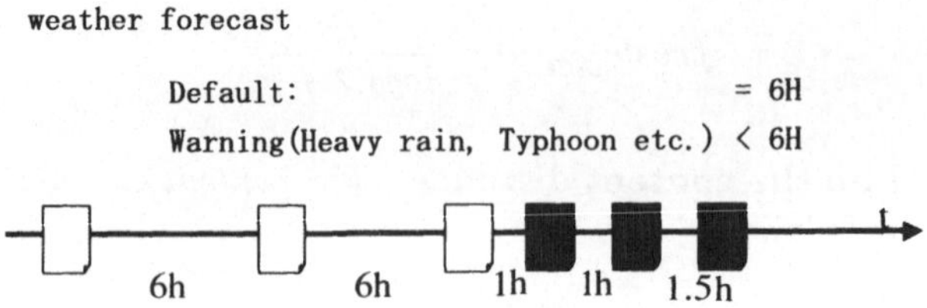

Figure 11: Example of update frequency

be also evaluated. Articles, which are quite similar to almost of the previously selected ones, would be also valuable. For example, when an incident happens, the series of the report articles would be sent continuously. That is, the articles are one of the *hottest* information at that time.

The *popularity* of the news article a can be also estimated by 1) the densimeter of similar articles of a in a retrospective scope, and 2) the time distance of a and similar articles in a retrospective scope (see Figure 10). In other words, if a has many similar articles in a retrospective scope and the time distance among them is small, the popularity of a is considered to be high. Thus, we have

$$pop(a) = e^{\lambda_1 k} + e^{-\lambda_2 t_d} \tag{4.10}$$

where $\lambda_1(> 0), \lambda_2(> 0)$ are the weight values. $k = m/n$ is the densimeter of similar articles, and t_d is the time distance between a and its similar articles, defined as follows:

$$t_d = \frac{1}{m} \sum_{i=1, b_i \in \omega}^{m} (t(a) - t(b_i)) \tag{4.11}$$

4.3 Urgency

Each real channel generally has its default update duration. However, in some cases, the duration becomes shorter than the default one because of urgency. For example, the channel of weather forecast would change their update frequency when typhoon warnings were announced (see Figure 11). In this case, the default update duration is six hours and the urgent update duration should be one hour. The urgency of channel c which article a belongs to is evaluated as follows:

$$freq(a) = e^{\lambda_1 \sigma_c} \tag{4.12}$$

$$\sigma_c = \frac{D_c - d_c}{d_c} \tag{4.13}$$

where σ_c is the ratio of update frequency of channel c, D_c is the default duration of channel c, d_c is the latest duration of channel c, and λ_1 is a weight value.

4.4 Filtering based on user profile and time-series fatures

Based on the user profile and the time-series features, we give a new filtering model for the news articles. In this model, the filter has three functions: (a) user profile matching, (b) channel update frequency monitoring, (c) popularity and freshness calculating. When the user profile for the filtered channel is q, the score of article a via channel c is calculated by following equation:

$$\begin{aligned} score(a) &= \alpha' * sim(a,q) + \beta' * freq(a) \\ &+\gamma' * (max(\mu * pop(a), \nu * fresh(a)) \end{aligned}$$

where sim(a, q) is the similarity between article a and user profile q, freq(a) is the urgency of channel c which article a belongs to, pop(a)is the popularity of a, fresh(a) is the freshness of article a. α', β', γ', μ, and ν are weight values for each term.

5 Concluding Remarks

In this paper, we focussed on the substantial aspects of data broadcasting systems. We noticed:

- Continuous delivery of content, in which update frequencies of their components may vary, may lead to some temporal inconsistency.
- Data content may be desirable to be reorganized into forms suitable for passive viewing.
- It is now a crucial issue how to obtain *fresh or popular* information from data broadcasting systems.

As for the first aspect, we introduced a version management mechanism to maintain the *temporal consistency* of delivered data by binary version trees for both of content updates and valid time updates. As for the second aspect, we described a way to view Web contents in a *passive manner* like TV-program contents. Our automatic transformation function from Web pages into TV-prigram-like contents raised an interesting data reorganization problem. That is, how can we tranform Web page content into synchronized data streams? The automatic transformation from Web pages into SMIL[11, 12] content will be also an interesting research issue. This question will need further research. Finally, as for the third aspect, we described our new information filtering mechanism for data broadcasting systems considering the *time-series features* of data, especially, the *freshness* and the *popularity* of data. Based on thess time-series features, we provide an new filtering method to obtain the fresh, popular and urgency information in data broadcasting systems.

Bibliography

1) PointCast, Pointcast network, http://www.pointcast.com.

2) Laura Lemay, "Official Marimba Guide to Castanet," Sams.net, (1997).

3) Kazutoshi Sumiya, Reiko Noda, and Katsumi Tanaka, "Hypermedia Broadcasting with Temporal Links," *Proc. of 9th International Conference on Database and Expert Systems Applications (DEXA '98)*, pp 176-185 (August 1998).

4) Kazutoshi Sumiya, Reiko Noda, and Katsumi Tanaka, "A Temporal Link Mechanism for Hypermedia Broadcasting (in Japanese)," *The Transaction of the Institute of Electronics, Information and Communication Engineers D-I*, J82-D-I(1), pp. 291-302 (1999).

5) Reiko Noda, Kazutoshi Sumiya, and Katsumi Tanaka, "A Valid-Time Based Delivery Model and Temporal Consistency Management for Hypermedia Broadcasting (in Japanese)," *IPSJ Transactions on Databases*, Vol.40, No.SIG8(TOD4), pp.126–140 (November 1999).

6) Taeko Hattori, Ikuo Sawanaka, Akiyo Nadamoto, and Katsumi Tanaka, "Discovering Synchronizable Regions and A Scripting Markup Language S-XML for Passive Web Browsing (in Japanese)," *IPSJ Technical Report of SIGDBS*, 2000-DBS-121-2, pp.9–16 (May 2000).

7) Katsumi Tanaka, Akiyo Nadamoto, Machiko Kusahara, Taeko Hattori, Hiroyuki Kondo, and Kazutoshi Sumiya, "Back to the TV: Information Visualization Interfaces Based on TV-Program Metaphors," *to appear in Proc. of the IEEE International Copnference on Multimedia and Expo (ICME2000)*, New York (July-August 2000).

8) TVML Consortium, http://www.strl.nhk.or.jp/TVML/

9) Qiang Ma, Hiroyuki Kondo, Kazutoshi Sumiya, and Katsumi Tanaka, "Virtual TV Channel: Filtering, Merging and Presenting Internet Broadcasting Channels," *Proc. of the ACM Digital Library Workshop on Organizing Web Space (WOWS)* (August 1999).

10) Qiang Ma, Kazutoshi Sumiya, and Katsumi Tanaka, "Information Filtering Based on Time-Series Features for Information Dissemination Systems (in Japanese)," *IPSJ Transactions on Databases*, Vol.41, To appear (2000).

11) Peiya Liu, "An Introduction to the Synchronized Multimedia Integration Language," *IEEE Multimedia*, Vol.5(No.4), pp. 84-88 (October-December 1998). http://www.computer.org/multimedia/mu1998/u4toc.htm.

12) World Wide Web Consortium(W3C), "Synchronized Multimedia Integration Language (SMIL) Boston Specification," http://www.w3.org/TR/smil-boston (November 1999).

5

Multimedia Database Systems Approaching Human Impression ("Kansei")

Yasushi Kiyoki

Keio University
Department of Environmental Information

ABSTRACT

In the design of multimedia database systems, one of the important issues is how to deal with "impression" of human beings. The concept of "human impression" includes several meanings on sensitive recognition, such as "human senses," "feelings," "sensitivity," "psychological reaction" and "physiological reaction".

In the field of database systems, the concept of multimedia databases is related to data definition and data retrieval with information on human impression for multimedia data, such as images, music and video. The important subject is to retrieve images and music dynamically according to the human impression. In this chapter, we describe some multimedia systems which manipulate information on human impression for defining and retrieving multimedia data.

In the research field of multimedia database systems, it is becoming important to deal with information on human impression for defining and retrieving media data according to impressions and senses of individual users.

The conceptual overview of a multimedia database system dealing with information on human impression is shown in this chapter. The essential functions for dealing with human impression are summerized. There are several research projects to realize these functions. In the design of the information on human impression for media data, the important issues are how to define and represent the metadata of media data and how to extract media data dynamically according to the user's impression and the data contents. Creation and manipulation methods of metadata for media data are summerized. Furthermore, some research projects for information on human impression are introduced.

1 Research Trends on "Kansei" Databases

1.1 Database systems for "Kansei" information

In the design of multimedia database systems, one of the important issues is how to deal with "Kansei" of human beings. The concept of "Kansei" includes several meanings on sensitive recognition, such as "human senses," "feelings," "sensitivity," "psychological reaction"and "physiological reaction" [17].

In the field of database systems, the concept of Kansei is related to data definition and data retrieval with Kansei information for multimedia data, such as images, music and video. The important subject is to retrieve images and music dynamically according to the user's impression given as Kansei information. We review some multimedia systems which manipulate Kansei information for defining and retrieving multimedia data.

As discussed in [17], the field of Kansei was originally introduced as the word "aesthetics" by Baumgrarten in 1750. The aesthetica of Baumgrarten had been established and succeeded by Kant with his ideological aesthetics. In the research field of multimedia database systems, it is becoming important to deal with Kansei information for human beings for defining and extracting media data according to impressions and senses of individual users.

The conceptual overview of a Kansei database system is shown in Fig. In the Kansei database system, the essential functions for dealing with Kansei in database systems can be summerized as follows:

(1) Defining Kansei information to media data (metadata definition for media data).

(2) Defining Kansei information for user's requests (metadata definition for user's requests(user's keywords) with Kansei information).

(3) Computing semantic correlations between Kansei information of media data and a user's request (media data retrieval subsystem with a correlation computation mechanism).

(4) Adapting retrieval results according individual variation and improving accuracy of the retrieval results by applying a learning mechanism to metadata (learning mechanism for metadata).

There are several research projects to realize these functions. In the design of the Kansei information for media data, the important issues are how to define and represent the metadata of media data and how to extract media data dynamically according to the user's impression and the data contents.

Creation and manipulation methods of metadata for media data have been summerized in [18, 19, 2]. Furthermore, some research projects for Kansei information have been established in academic fields [17]. As one of Kansei research projects, "modeling the evaluation structure of Kansei" has started in 1997 [17]. Multimedia database subjects related to kansei information retrieval are promoted in this project.

There are many approaches to media data retrieval. Two major approaches are direct retrieval using partial pattern matching and indirect retrieval using abstract information of images.

Several multimedia database systems for Kansei information retrieval have been proposed. The pictorial information server systems, named TRADEMARK and Electrical Art Gallery Art Museum, have been proposed to perform picture retrieval using query-by-visual-example and query-by-subjective-descriptions [3]. The query-by-visual-example provides sketch retrieval facility to find similar pictorial data without textual information. The query-by-subjective-description provides a facility for a user to show his own emotional representations to find pictorial data which is appropriate to his subjective interpretation automatically evaluating the content of the pictorial data. Those systems have been implemented with several functions for computing correlations between the user's request and retrieval candidate pictorial data.

As one of the database systems dealing with Kansei information, we have introduced a semantic associative search system for images [5]. The semantic associative search system realizes image data retrieval by receiving keywords representing the user's impression and the image's contents. This system provides several functions for performing the semantic associative search for images by using the metadata representing the features of images. These functions are realized by using the mathematical model of meaning [4, 5]. The mathematical model of meaning provides semantic functions for computing specific meanings of keywords which are used for retrieving images unambiguously and dynamically. The main feature of this model is that the semantic associative search is performed in the orthogonal semantic space. This space is created for dynamically computing semantic equivalence or similarity between the metadata items of the images and keywords.

A structure of Kansei information database has been proposed for supporting design processes by constructing static and dynamic image information of human motions and positions [15]. The main purpose of this project is to structure information regarding movement and posture of human bodies as databases for supporting tools for design ideas. The movement of hands(animated, still images) and operation sound (effects) can be included as the contents of the database. Currently, images of hand motions operating equipments are stored in databases and those data are manipulated by the senses of sight, touching, and listening.

A learning mechanism is very important for database systems dealing with Kansei information to adapt retrieval results according to individual variation and improving accuracy of the retrieval results. Such database systems might not always select accurate and appropriate data items from databases, because the judgement of accuracy for the retrieval results is strongly dependent on individual variation. In the learning, if inappropriate retrieval results for a request are extracted by the system, accurate data items which must be the retrieval results are specified as suggestions. Then, the learning mechanism is applied to the system to extract the appropriate retrieval results in subsequent requests.

Several approaches for adapting retrieval results to the user's impression have been proposed. In [11], in the framework of query-by-visual-example, individual variations of users are reflected by adapting individual user's information to database contents. This method is based on the computations of correlations between images data and user's individual data which are represented in vectors which consist of color elements.

In [8], the learning mechanism has been proposed for applying it to the semantic associative search system. In this learning, if inappropriate retrieval results for a request are extracted by the semantic associative search, accurate data items which must be the retrieval results are specified as suggestions. Then, the learning mechanism is applied to the semantic associative search system to extract the appropriate retrieval results in subsequent requests.

1.2 A semantic associative search system for Kansei information retrieval

As an associative search system related to Kansei information, we review a semantic associative search system for image databases dealing with Kansei information[4, 5]. The mathematical model of meaning has been designed for realizing the semantic associative search with semantic computation machinery for context recognition[4, 5]. This model can be applied for retrieving multimedia information, such as images and music, with Kansei information. This model can be applied to extract images by giving the context words which represent the user's impression and contents of the images.

Images and contexts are characterized by the specific features(words) as Kansei information and those features are represented as vectors. Those vectors are named "metadata items for images" and "metadata items for keywords."The important feature of this model is that semantic similarity between a given context and images is measured by using the following mathematical operations on the orthogonal semantic space.

The mathematical model of meaning consists of:

1) A set of m words is given, and each word is characterized by n features. That is, m by n matrix is given as the data matrix M.

2) The correlation matrix of M with respect to the n features is constructed. Then, the eigenvalue decomposition of the correlation matrix is computed and the eigenvectors are normalized. The orthogonal semantic space is created as the span of the eigenvectors which correspond to nonzero eigenvalues.

3) Images are characterized by the specific features(words) which correspond to the n features in the step 1), and the metadata items for the images are represented as vectors with the n elements. the metadata items for keywords are also characterized by the same features and represented as vectors.

4) The metadata items for images and keywords are mapped into the orthogonal semantic space by computing the Fourier expansion for the vectors.

5) A set of all the projections from the orthogonal semantic space to the invariant subspaces (eigen spaces) is defined. Each subspace represents a phase of meaning, and it corresponds to a context or situation.

6) A subspace of the orthogonal semantic space is selected according to the user's impression, which is given as a context representing Kansei information with a sequence of context words.

7) The most correlated image (semantically closest image) to the given context as the user's impression (Kansei information) is extracted in the selected subspace.

The semantic associative search system selects appropriate images for user's requests with Kansei information by using metadata items and basic functions. This system consists of the following subsystems:

(1) Image Selection Subsystem: This subsystem supports the facilities for selecting appropriate images by using the mathematical model of meaning. Three methods are provided for representing the metadata items for images. This subsystem maps the metadata items in the semantic space created in the step 2). When the context is given as the Kansei information, this subsystem selects the most correlated image to the context.

(2) Metadatabase Management Subsystem: This subsystem supports the facilities for keeping metadata consistent in the orthogonal semantic space.

(3) Metadata Acquisition Subsystem: This subsystem supports the facilities for acquiring metadata from the database storing the source images by receiving user's requests with Kansei information.

1.3 Basic functions and metadata for images

The metadatabase system is used to extract image dta items corresponding to context words which represent the user's impression and the image's contents. For example, the context words "powerful" and "strong" are given, the image with the impression corresponding to these context words is extracted. Each metadata item of images is mapped in the orthonormal semantic space. This space is referred to as "orthogonal metadata space" or "metadata space." The MMM is used to create the orthogonal metadata space. By this orthogonalization, we can define appropriate metric for computing relationships between metadata items for images and context representation. The MMM gives the machinery for extracting the associated information to the context.

Three types of metadata are used.

(1) *Metadata for space creation*: These metadata items are used for the creation of the orthogonal metadata space, which is used as a space for semantic image retrieval.

(2) *Metadata for images*: These metadata items are the candidates for semantic image retrieval.

(3) *Metadata for context representation*: These metadata items are used as context words for representing the user's imagination and the image's contents.

The basic functions and metadata structures are summarized as follows:

1. Creation of metadata space:

   To provide the function of semantic associative search, basic information on m data items ("metadata for space creation") is given in the form of a matrix. Each metadata item is given independently one another. No relationship between the metadata items need to be described. The information of each data item is represented by its features. The m basic metadata items is given in the form of an m by n matrix M where each metadata item is characterized by n features. By using M, the orthogonal space is created as the metadata space $\mathcal{MDS}$. These metadata items are determined as mentioned in the following section.

2. Representation of metadata for images in n-dimensional vectors

   Each metadata item for images is represented in the n-dimensional vector whose elements correspond to n features. The metadata items for images become the candidates for the semantic associative search. Furthermore, each of the context words, which are used to represent the user's impression and the image's contents in semantic image retrieval, is also represented in the n-dimensional vector.

3. Mapping data items into the metadata space $\mathcal{MDS}$.

   Metadata items (metadata for space creation, metadata for images and metadata for context representation) which are represented in n-dimensional vectors are mapped into the orthogonal metadata space. Those data items are used as context words and target image data items which are extracted according to users' requests.

4. Semantic associative search

   When a sequence of context words which determine the user's impression and the image's contents is given, the images corresponding to the context are extracted from a set of retrieval candidate images in the metadata space.

1.4 A creation method of the metadata space

We introduce an implementation method for the creation of the $\mathcal{MDS}$.

The procedure for the creation of the $\mathcal{MDS}$ is as follows:

1. To create the data matrix M, we can use "General Basic English Dictionary[16]" in which 850 basic words are used to explain each English definition. Those 850 basic words are used as features, that is, they are used for characterizing metadata as the features corresponding to the columns in the matrix M. The 2,000 data items are used as "metadata for space creation." Those metadata items are used as the basic words in the English dictionary "Longman Dictionary of Contemporary English [12]." Each metadata item for space creation corresponds to a row of the matrix M. In the setting of a row of the matrix

M, each column corresponding to the features which appear in the explanation of the data item is set to the value "1". If the feature is used as a negative meaning, the column corresponding to it is set to the value "-1". And, the other columns are set to the value "0". This process is performed for each of 2000 metadata items. And then, each column of the matrix is normalized by the 2-norm to create the matrix M.

2. By using this matrix M, the $\mathcal{MDS}$ is created as the orthogonal space. The creation method of the orthogonal space is described in Section 3.1. This space represents the semantic space for computing contexts and meanings of the metadata items.

To automatically create the data matrix M from the dictionary, several filters are used, which remove unnecessary elements (words), such as articles and pronouns, and transform conjugations and inflections of words to the infinitives. Those elements are removed from the features characterizing each data item. The unnecessary words are not used as features in the data matrix M.

(filter-1) This filter eliminates the unnecessary elements, such as articles and pronouns.

(filter-2) This filter transforms conjugations and inflections to the infinitives.

(filter-3) This filter transforms the large characters to the small ones.

(filter-4) This filter transforms clipped form words to the corresponding original words.

(filter-5) The rows of the matrix M are created for each data item by using the filtered features which characterize the data item.

Each metadata item (metadata item for images, metadata item for context representation) is mapped into the metadata space, by computing the Fourier expansion for the n-dimensional vector representing the metadata item itself. These metadata items are defined as metadata by using the n features. These metadata items are used as context words and metadata items for retrieval candidate images.

2 Creation of a Metadata Space and Basic Functions

In this section, we introduce a creation method of the metadata space $\mathcal{MDS}$ for systematically storing metadata and for implementing the semantic associative search for images[4, 5].

2.1 Creation of a metadata space

The semantic associative search for images is created by using our mathematical model of meaning [4, 7]. For the metadata items for space creation, a data matrix M is created. When m data items for space creation are given, each data item is characterized by n features $(f_1, f_2, \cdots, f_n)$. For given $\mathbf{d}_i (i = 1, \cdots, m)$, the data matrix M is defined as the $m \times n$ matrix whose i-th row is $\mathbf{d}_i$. Then, each column of the matrix is normalized by the 2-norm in order to create the matrix M.

Figure 1 shows the matrix M. That is $M = (\mathbf{d}_1, \mathbf{d}_2, \mathbf{d}_3, \cdots, \mathbf{d}_n)^T$.

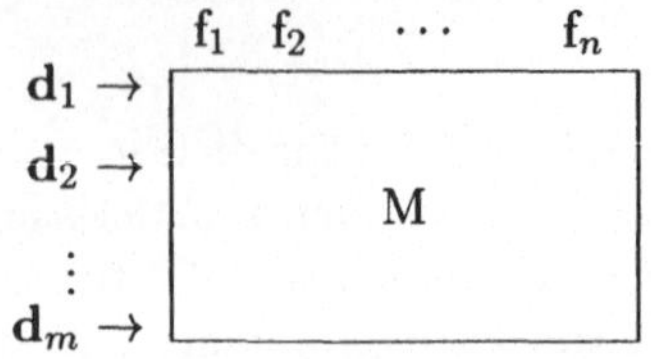

Figure 1: Representation of metadata items by matrix M

1. The correlation matrix $M^T M$ of M is computed, where M^T represents the transpose of M.

2. The eigenvalue decomposition of $M^T M$ is computed.

$$M^T M = Q \begin{pmatrix} \lambda_1 & & & & \\ & \ddots & & & \\ & & \lambda_\nu & & \\ & & & 0 & \\ & & & & \ddots 0 \end{pmatrix} Q^T,$$

   $0 \leq \nu \leq n$.
   The orthogonal matrix Q is defined by

$$Q = (\mathbf{q}_1, \mathbf{q}_2, \cdots, \mathbf{q}_n)^T,$$

   where $\mathbf{q}_i$'s are the normalized eigenvectors of $M^T M$. We call the eigenvectors "semantic elements" hereafter. Here, all the eigenvalues are real and all the eigenvectors are mutually orthogonal because the matrix $M^T M$ is symmetric.

3. Defining the metadata space $\mathcal{MDS}$.

$$\mathcal{MDS} := span(\mathbf{q}_1, \mathbf{q}_2, \cdots, \mathbf{q}_\nu),$$

   which is a linear space generated by linear combinations of $\{\mathbf{q}_1, \cdots, \mathbf{q}_\nu\}$. We note that $\{\mathbf{q}_1, \cdots, \mathbf{q}_\nu\}$ is an orthonormal basis of $\mathcal{MDS}$.

2.2 The set of the semantic projections Π_ν

The projection P_{λ_i} is defined as follows:
$P_{\lambda_i} \stackrel{\mathrm{d}}{\Longleftrightarrow}$ Projection to the eigenspace corresponding to the eigenvalue λ_i,
i.e. $P_{\lambda_i} : \mathcal{MDS} \rightarrow span(\mathbf{q}_i)$.
The set of the semantic projections Π_ν is defined as follows:

$$\Pi_\nu :=$$

$$\begin{gathered} \{\, 0\ ,\ P_{\lambda_1}, P_{\lambda_2}, \cdots, P_{\lambda_\nu}, \\ P_{\lambda_1} + P_{\lambda_2}, P_{\lambda_1} + P_{\lambda_3}, \cdots, P_{\lambda_{\nu-1}} + P_{\lambda_\nu}, \\ \vdots \\ P_{\lambda_1} + P_{\lambda_2} + \cdots + P_{\lambda_\nu} \}. \end{gathered}$$

The number of the elements of Π_ν is 2^ν, and accordingly it implies that 2^ν different contexts can be expressed by this formulation.

2.3 Semantic operator

The correlations between each context word and each semantic element are computed by this process. The context word is used to represent the user's impression and the image's contents for images to be extracted. A sequence

$$s_\ell = (\mathbf{u}_1, \mathbf{u}_2, \cdots, \mathbf{u}_\ell)$$

of ℓ context words and a positive real number $0 < \varepsilon_s < 1$ are given, the semantic operator S_p constitutes a semantic projection $P_{\varepsilon_s}(s_\ell)$, according to the context. That is,

$$S_p : T_\ell \longmapsto \Pi_\nu$$

where T_ℓ is the set of sequences of ℓ words and $T_\ell \ni s_\ell$, $\Pi_\nu \ni P_{\varepsilon_s}(s_\ell)$. Note that the set $\{\mathbf{u}_1, \mathbf{u}_2, \cdots, \mathbf{u}_\ell\}$ must be a subset of the words defined in the matrix M.

The constitution of the operator S_p consists of the following processes:

1. Fourier expansion of $\mathbf{u}_i (i = 1, 2, \cdots, \ell)$ is computed as the inner product of $\mathbf{u}_i$ and $\mathbf{q}_j$ u_{ij}, i.e.

   $$u_{ij} := (\mathbf{u}_i, \mathbf{q}_j)\ ,\ for\ j = 1, 2, \cdots, \nu.$$

   $\widehat{\mathbf{u}}_i \in \mathcal{I}$ is defined as

   $$\widehat{\mathbf{u}}_i := (u_{i1}, u_{i2}, \cdots, u_{i\nu}).$$

   This is the mapping of the context word $\mathbf{u}_i$ to the $\mathcal{MDS}$.

2. The semantic center $\mathbf{G}^{+}(s_\ell)$ of the sequence s_ℓ is computed as,

$$\mathbf{G}^{+}(s_\ell) := \frac{\left(\sum_{i=1}^{\ell} u_{i1}, \cdots, \sum_{i=1}^{\ell} u_{i\nu}\right)}{\| \left(\sum_{i=1}^{\ell} u_{i1}, \cdots, \sum_{i=1}^{\ell} u_{i\nu}\right) \|_\infty},$$

where $\| \cdot \|_\infty$ denotes infinity norm.

3. The semantic projection $P_{\varepsilon_s}(s_\ell)$ is determined as,

$$P_{\varepsilon_s}(s_\ell) := \sum_{i \in \Lambda_{\varepsilon_s}} P_{\lambda_i} \quad \in \Pi_\nu,$$

where $\Lambda_{\varepsilon_s} := \{\, i \mid |(\mathbf{G}^{+}(s_\ell))_i \mid > \varepsilon_s\}$.

2.4 Function for Semantic Image Search

We introduce a function to measure the similarity between images and context words. The function measures the quantity of association or correlation between context words and the candidate images. We also introduce a dynamic metric depending on the context between two images.

2.4.1 Function to measure the association

The function measures the association between context words and the candidate images. Suppose a sequence of associate context words is given to search an image, e.g. { dynamic, powerful }. We can regard the context words as the words forming the context s_ℓ. We can specify some subspace by using the semantic operator with weights c_j's which are given by

$$c_j(s_\ell) := \frac{\sum_{i=1}^{\ell} u_{ij}}{\| \left(\sum_{i=1}^{\ell} u_{i1}, \cdots, \sum_{i=1}^{\ell} u_{i\nu}\right) \|_\infty},$$

$$j \in \Lambda_{\varepsilon_s}.$$

Since the norm of the image, which can be calculated from the metadata of the image, reflects the correlation between the image and the semantic elements included in the selected subspace, we may use it as a measure for the association between the context words and the image data.

We introduce a function for computing the norm of the image. The function $\bar{\eta}_0\,(\mathbf{x}; s_\ell)$ is defined as follows:

$$\bar{\eta}_0\,(\mathbf{x}; s_\ell) = \frac{\sqrt{\sum_{j \in \Lambda_{\varepsilon_s} \cap \mathcal{S}} \{c_j(s_\ell) x_j\}^2}}{\|\mathbf{x}\|_2},$$

where the set $\mathcal{S}$ is defined by $\mathcal{S} = \{i | sign(c_i(s_\ell) = sign(x_i)\}$.

In this function, we eliminate the effect of the negative correlation. We note that the sum in the numerator of this function is sought over the selected semantic subspace from the context s_ℓ, while the norm in the denominator is sought over the whole metadata space $\mathcal{MDS}$.

2.4.2 Dynamic metric

We introduce a dynamic metric between the image data items, according to a context. Since each image data item can be represented as a vector via the union operator $\oplus$ defined in Section 4.2, we can utilize the metric, which we defined for two distinct words in [4, 5, 7], to compute the similarity between metadata items of images. The dynamic metric $\rho(\mathbf{x}, \mathbf{y}; s_\ell)$ for $\mathbf{x}, \mathbf{y} \in \mathcal{MDS}$ is introduced to compute the similarity between metadata items for images.

The metric $\rho(\mathbf{x}, \mathbf{y}; s_\ell)$ to compute the similarity between metadata items $\mathbf{x}$, $\mathbf{y}$ of two images in the given context s_ℓ is defined as follows:

$$\rho(\mathbf{x}, \mathbf{y}; s_\ell) = \sqrt{\sum_{j \in \Lambda_{\varepsilon_s}} \{c_j(s_\ell)(x_j - y_j)\}^2}.$$

This metric, because of the presence of dynamic weights c_j's, can faithfully reflect the change of the context.

3 Semantic Associative Search for Metadata for Images

The proposed system realizes the semantic associative search for metadata items for images.

The basic function of the semantic associative search is provided for context-dependent interpretation. This function performs the selection of the semantic subspace from the metadata space. When a sequence s_ℓ of context words for determining a context is given to the system, the selection of the semantic subspace is performed. This selection corresponds to the recognition of the context, which is defined by the given context words. The selected semantic subspace corresponds to a given context. The metadata item for the most correlated image to the context in the selected semantic subspace is extracted from the specified image data item set $\mathcal{W}$. By using the function defined in Section 2.4.2, the semantic associative search is performed by the following procedure:

1. When a sequence s_ℓ of the context words for determining a context(the user's impression and the image's contents) are given, the Fourier expansion is computed for each context word, and the Fourier coefficients of these words with respect to each semantic element are obtained. This corresponds to seeking the correlation between each context word and each semantic element.

2. The values of the Fourier coefficients for each semantic element are summed up to find the correlation between the given context words and each semantic element.

3. If the sum obtained in the step 2 in terms of each semantic element is greater than a given threshold ε_s, the semantic element is employed to form the seman-

tic subspace $P_{\varepsilon_s}(s_\ell)\mathcal{MDS}$. This corresponds to recognizing the context which is determined by the given context words.

4. By using the function $\bar{\eta}_0(\mathbf{x}; s_\ell)$, the metadata item for the image with the maximum norm is selected among the candidate metadata items for images in $\mathcal{W}$ in the selected semantic subspace. This corresponds to finding the image with the greatest association to the given context words from $\mathcal{W}$.

4 Creation methods of metadata for images

We present three methods for creating metadata for images.

In [2], Kashyap et al. have clearly identified and classified various metadata for digital media into three basic categories: *content-dependent metadata*, *content-descriptive metadata* and *content-independent metadata*. The metadata for images which is used in our semantic associative search is categorized in the content-descriptive metadata, because the metadata is associated with the original image without being extracted directly from the image contents themselves. Furthermore, in [2], the content-descriptive metadata is classified into two categories: *domain-dependent metadata* and *domain-independent metadata*.

In the following Method-1 and Method-2, the metadata for images is categorized into *domain-dependent* because the metadata is extracted from domain-specific concepts, which are used as a basis for the determination of the metadata itself. That is, the metadata type used in these methods is categorized as the *content-descriptive domain-dependent metadata*.

In Method-3, metadata for images is extracted from their color components which are used to characterize image features in an experimental psychology model of correlating colors and their impression words. This type of metadata is categorized as the *content-descriptive domain-independent metadata*.

4.1 Method-1

Each image is explained by using the n features which are used in the creation of the data matrix M. In this explanation, the impression or the content of the image is represented by using these features as the metadata for the image. As the result, each image is represented as n-dimensional vector in which the non-zero value is assigned to the corresponding elements of the vector to these features.

The image P is explained and defined by using some of the words which are used in the n features. Then, the image is represented as an n-dimensional vector.

$$P = (w_{i1}, w_{i2}, \ldots, w_{in}).$$

Each metadata item is mapped into the metadata space by computing the Fourier expansion for the vector corresponding to the image data item itself.

4.2 Method-2

The image P is represented in t impression words $\mathbf{o}_1, \mathbf{o}_2, \ldots, \mathbf{o}_t$, where each impression word is defined as a t dimensional vector:

$$\mathbf{o}_i = (o_{i1}, o_{i2}, \ldots, o_{in}),$$

which is characterized by n specific features.

Namely, we define the image P as the collection of t impression words which represent the image.

$$P = \{\mathbf{o}_1, \mathbf{o}_2, \ldots, \mathbf{o}_t\}.$$

Moreover, we define the operator union $\bigoplus$ of impression words $\mathbf{o}_1, \mathbf{o}_2, \ldots, \mathbf{o}_t$, to represent the metadata for the image P as a vector as follows:

$$\bigoplus_{i=1}^{t} \mathbf{o}_i \equiv (\text{sign}(o_{\ell_1 1}) \max_{1 \le i \le t} |o_{i1}|, \text{sign}(o_{\ell_2 2}) \max_{1 \le i \le t} |o_{i2}|,$$

$$\ldots, \text{sign}(o_{\ell_n t}) \max_{1 \le i \le t} |o_{in}|)$$

where $\text{sign}(a)$ represents the sign (plus or minus) of "a" and $\ell_k, k = 1, \ldots, n$, represents the index which gives the maximum, that is:

$$\max_{1 \le i \le t} |o_{ik}| = |o_{\ell_k k}|.$$

4.3 Method-3

In this method, the metadata for images is automatically and indirectly extracted from image data items themselves. Color is known as the dominant factor which affects the impression of images[10, 11, 1]. We use color to derive impressions of images.

The basic idea of this method is to describe both images and impression words in the notion of *color* and compute correlations between images and words. Color used in this method is represented in the Munsell color system as it is more familiar to the human perception. Additionally, the color names defined by ISCC(Inter-Society Color Council) and NIST(National Institute of Standards and Technology) are used to describe both images and impression words in the notion of color. By taking the correlations between images and impression words, we can obtain the suitable words which describe the impressions of images. The metadata for images is computed from the obtained impression words by the previously defined union operator $\bigoplus$.

5 Examples of Creating Metadata

5.1 Method-1

To create the metadata as vectors for images by using the 850 basic words of "General Basic English Dictionary," the designer of metadata looks at an image and checks features that correspond to the image. If the feature corresponds to the image, the value 1.0 is put for that feature, if it does not correspond to the feature, the value 0.0 is put, and if it negates the image, the value -1.0 is put. Although the cost is very expensive, the simplest way is to check for 850 features one by one for each image. A vector is created for each image, and mapped into the $\mathcal{MDS}$.

5.2 Method-2

As the previous method, the same features from "General Basic English Dictionary" are used. This method creates metadata by giving impressions of the original image or referring to objects composing it. The impression words or object names which are extracted from the image are referred to from "General Basic English Dictionary," and the explanatory words for each impression word or object name are checked as features, and the value for each feature in the vector corresponding to the impression word or object name is set. Then, the vector corresponding to the image is created from the vectors corresponding to the impression words or object names by the union operator defined in Section 4.2.

5.3 Method-3

In this method, metadata for an image is created by referring to color components. Digital images, which usually represented in the RGB color system, are transformed to color vectors in the Munsell color space by using the MTM(Mathematical Transformation to Munsell)[14, 13].

Scaler values of color vectors for corresponding color are defined in the range of 0.0 and 1.0 according to given rules. One of the rules which we have used defines each value by referring occupancy of colors in images.

The difficulty of this method is to define the association between the colors and impression words. That is, how to create the descriptive metadata for context words. To solve this difficulty, we referred to the results from the field of the experimental psychology. Many word association tests have been done to make clear the relation between colors and psychological effects.

6 Automatic extraction of Kansei information from images

In this section, we consider a method for automatically extracting Kansei information from images. The metadata for images is created by using the image data items themselves. It is implemented by using image processing facilities which are studied in the field of image processing.

If we consider the case that the impression which we obtain from an image can be estimated from the elements composing the image, we can derive the following formula:

$$P = \sum_{i=1}^{N} a_i I_i, \tag{5.1}$$

where P is an impression vector of an image itself, N is a number of elements composing the image, a_i is some kind of coefficient for element i in the image, and I_i is an impression vector of the element i. Each impression vector is expressed as follows:

$$I_j = (w_{j_1}, w_{j_2}, ..., w_{j_m}), \tag{5.2}$$

where w_{j_i} is a word expressing an impression, and m is the number of features for expressing the impression. The actual value of w_{j_i} is the weight of that word.

We consider the case that the colors used in the image can derive the impression of the image, that is, the color is the dominant factor which decides the impression of the image[1, 11, 9]. Images and colors are described by expression words such as 'warm', 'bright' and so on.

The coefficient a_i of the formula is the percentage of the region of specific color, and I_i is the impression vector of that color, that is, its elements correspond to colors. The impression vector of the image P, that is, the metadata for the image, can be expressed as follows:

$$P = (c_{p_1}, c_{p_2}, ..., c_{p_m}), \tag{5.3}$$

where c_{p_i} is a color used on the image. Actual value of c_{p_i} is the weight of that color in the image. In the case that only the area size of each color is considered, c_{p_i} should be calculated by the following formula:

$$c_{p_i} = k \frac{A_{c_{p_i}}}{A}, \tag{5.4}$$

where k is a constant value, A_{c_i} is the unit area of a color c_i, and A is the unit area of the whole image. Note that c_i must be within $[0, 1]$. Similarly, the word w_i, metadata for a keyword, is expressed as follows:

$$w_i = (c_1, c_2, ..., c_m). \tag{5.5}$$

As the result, the distance between images and words can be measured.

It is possible to use the Munsell color system to express colors, which is much more familiar to the human sense than the RGB color system.

We can use the results of psychological experiments, as many word association tests have been done on the relation between colors and psychological effects, e.g. showing a single color and asking the reminded words. According to the results, there seems to be an association between words and colors. For instance, from the color 'strong orange', the word 'warm' is likely to be associated with it. Naturally, the association between colors and words is not one-to-one correspondence, but many-to-many.

For example, the color names defined by ISCC and NBS can be used as the features. The creation of the metadata for images can be done automatically, as the color of the image in the CIE color system can be obtained easily and can be transformed to the Munsell color system. The creation of the metadata can be done automatically.

However, the difficulty in automatic extraction is how to correspond the colors with explanatory keywords, that is, how to create the metadata for keywords. To solve this difficulty, the results of psychological experiments can be used, because many word association tests had been done on the relation between colors and psychological effects, e.g. showing a single color and asking for the reminded words.

7 Conclusion

In this chapter, we have reviewed several database systems dealing with Kansei information for extracting media data according to the user's impression and the image's contents. Those system provide functions for defining and retrieving images with Kansei information. Those functions are realized by computing correlations between Kansei information of media data and a user's request represented with Kansei information.

In this field, the development for learning mechanisms is important for supporting adaptivity to individual variations in Kansei. The implementation of automatic extraction mechanisms for Kansei information is also very important to realize actual Kansei database system environments.

We have introduced new methodology for retrieving image data according to the user's impression and the image's contents. We have presented functions and metadata for performing semantic associative search for images. The functions are realized on the basis of mathematical model of meaning.

For the creation of the metadata for images, we have introduced three methods (Method-1, Method-2 and Method-3). The metadata created by those methods is categorized in the type of *content-descriptive metadata* according to the metadata classification for digital media presented in [2]. Furthermore, the metadata created by the first two methods, Method-1 and Method-2, is categorized into the content-descriptive domain-dependent metadata, and the metadata by the third method is classified as the type of content-descriptive domain-independent metadata.

We have implemented the semantic associative search system to clarify its feasibility and effectiveness. Currently, we are designing a learning mechanism to adapt metadata for context representation and images according to the individual variation. The learning is a significant mechanism for semantic associative search, because the judgment of accuracy for the retrieval results might be dependent on individuals.

We will use this system for realizing a multimedia metadatabase environment. As our future work, we will extend this system to support multimedia data retrieval for video and audio data. This system will be integrated in a distributed multimedia database environment[6, 7].

Acknowledgments

Special thanks to Prof. Takashi Kitagawa (Univ. of Tsukuba) for his valuable comments and suggestions on this work.

Bibliography

1) Chijiiwa, H., *"Color Science"*, Fukumura Printing Co., 1983.

2) Kashyap, V., Shah, K., Sheth, A., *"Metadata for Building the Multimedia Patch Quilt,"* V.S. Subrahamanian, S. Jajodia, eds., Multimedia Database Systems, pp.297–319, 1996.

3) Kato, T. *"Understanding Kansei - A cognitive aspect of human computer interaction -,"* Report of modeling the evaluation structure of KANSEI 1997, pp. 99-108, Akira Harada (eds.).

4) Kitagawa, T. and Kiyoki, Y., *"A Mathematical Model of Meaning and its Application to Multidatabase Systems,"* Proc. 3rd IEEE International Workshop on Research Issues on Data Engineering: Interoperability in Multidatabase Systems, pp.130–135, April 1993.

5) Kiyoki, Y., Kitagawa, T., and Hayama, T., *"A Metadatabase System for Semantic Image Search by a Mathematical Model of Meaning,"* ACM SIGMOD Record (Special issue on metadata for digital media), W. Klaus, A. Sheth, eds., Vol.23, No. 4, pp.34–41, Dec. 1994.

6) Kiyoki, Y. and Hayama, T., *"The Design and Implementation of a Distributed System Architecture for Multimedia Databases,"* Proc. 47th Conference of International Federation for Information and Documentation, pp.374–379, Oct. 1994.

7) Kiyoki, Y., Kitagawa, T. and Hitomi, Y., *"A Fundamental Framework for Realizing Semantic Interoperability in a Multidatabase Environment,"* Journal of Integrated Computer-Aided Engineering, Vol.2, No.1(Special Issue on Multidatabase and Interoperable Systems), pp.3–20, John Wiley & Sons, Jan. 1995.

8) Kiyoki, Y., Miyagawa, A. and Kitagawa, T., *"A multiple view mechanism with semantic learning for multidatabase environments,"* Information Modelling and Knowledge Bases (IOS Press), Vol. IX, May, 1998.

9) Kiyoki, Y., Kitagawa, T., and Hayama, T., *"A Metadatabase System for Semantic Image Search by a Mathematical Model of Meaning,"* Multimedia Data Management – using metadata to integrate and apply digital media –," McGrawHill(book), A. Sheth and W. Klas(editors), Chapter 7, March 1998.

10) Kurita, T., Kato, T., *"Learning A Cognitive Schema for Multimedia Indexing - Experiments on Image Database-,"* Technical Reports of IEICE, DE93–3, May 1993.

11) Kurita, T., Kato, T., Fukuda, I., Sakakura, A., *"Sense Retrieval on a Image Database of Full Color Paintings,"* Journal of Japan Information Processing Society, Vol.33, No.11, pp.1373–1383, Nov.1992.

12) *"Longman Dictionary of Contemporary English,"* Longman, 1987.

13) Miyahara, M. and Yoshida, Y., *"Mathematical Transformation of (R,G,B) Color Data to Munsell (H,V,C) Color Data"*, SPIE's Visual Communications and Image Processing '88, Vol.1001, No.118, pp.650–657, Nov. 1988.

14) Newhall,S.M., Nickerson,D. and Judd, D.B., *"Final Report of the O.S.A Subcommittee on the Spacing of the Munsell Colors"*, Journal of the Optical Society of America, Vol.33, No.7, pp.485–418 Jul. 1943.

15) Okazaki, A., *"KANSEI manipulation in design process-1,"* Report of modeling the evaluation structure of KANSEI 1997, pp. 63-71, Akira Harada (eds.).

16) Ogden, C.K., *"The General Basic English Dictionary,"* Evans Brothers Limited, 1940.

17) *"Report of modeling the evaluation structure of KANSEI 1997* Akira Harada (eds.).

18) *"Multimedia Data Management - Using Metadata to Integrate and Apply Digital Media,"* Amit Sheth, Wolfgang Klas (eds.), MacGraw-hill.

19) *"Special issue on metadata for digital media,"* ACM SIGMOD Record, W. Klaus, A. Sheth, eds., Vol.23, No. 4, Dec. 1994.

6

Heijo — A Video Database System for Retrieving Semantically Coherent Video Information

Shunsuke Uemura
Masatoshi Yoshikawa
Toshiyuki Amagasa

ABSTRACT

This chapter describes a research and development project for a video database system "Heijo." Retrieving meaningful video fragments from a video database is a challenging issue. Heijo can process such a query by combining simple indices constructed on each video data. To this end, three kinds of operations on time intervals, namely, intersection, union, and joint operations and algorithms for them are proposed. In addition, five kinds of time axes of video data and categorization of them based on the time axes are investigated so that we can distinguish video data by their contents. Finally, an overview of Heijo system is shown.

1 Introduction

Rapid progress in computer science and network technologies are realizing world wide information resources. Not only the volumes of those information resources are increasing, but also the kind of media being digitized is changing, namely from fairly simple character strings to such entities as those directly appeal to the senses of humans like pictures and sounds. Next-generation database systems should be able to integrate and manipulate those digital media.

This chapter describes a research and development project for video database systems or database systems for video media. We use the terminology "video" as a generic name for motion picture and its accompanying sound. One of the most important research topics for video database system is how to capture semantically unified and meaningful video fragment or semantically coherent video information. Traditional data model for video consists of "frame (still picture)", "cut (a series of frames)", and "scene (a series of cuts)", Frame, cut, and scene are all somewhat physical concepts. It is difficult to get meaningful video fragment from those concepts. Searching for semantically coherent video information can never be achieved with motion pictures only, or with audio media alone. Cooperative processing of simple indices is required. In addition, handling multiple scenes that occurred at the same time, but recorded on different videos, cannot be done effectively with conventional (individual) indexing method. Here, again "cooperative" processing of indices becomes important.

In this chapter, we presume simple indexing for each single media is available. We then discuss a method that finds out semantically coherent video objects while using this indexing in a practical manner. By "video object" we mean "semantically coherent video information (picture and sound)". A video object is defined by a time interval i.e. by a pair of ⟨*start time*⟩ and ⟨*end time*⟩ [7].

We propose a "joint" operation for time intervals, introduce its applications, and report an experimental implementation of a prototype video database system called the "Heijo" system.

Related research works on video database systems include the OVID system [8] and Video Algebra [9, 10]. The OVID system uses a video object data model, and combines multimedia data with an object-oriented database. Video Algebra is the proposal of a set of operation for carrying out such operations as connection of videos, parallel and simultaneous "play", or repeated "play". They are mainly oriented toward the playing of video, and database-type operations are never considered. A well known research on searching for semantically coherent objects from among a flood of video information is InforMedia. InforMedia, advocated a search for necessary information by the quoting of all necessary means [3, 4].

As for time interval operations, there is the pioneering work by Allen who classified time-based relations [1]. Since then, there have been proposals on time interval relations related to sets of multiple time intervals [5, 6], research to introduce empty time intervals [7, 2], and research on handling all multimedia objects in a unified way through time intervals [7].

Our research starts from a position close to InforMedia, but we discuss an original method that searches for meaningful video fragments (semantically coherent video object) by combining simple indices, and propose a new time interval operation for that purpose. In addition, we take up the problem of operations on a number of different media over the same time axis, and discuss how we can apply the proposed video operation even to relations of the multiple media whereby separate indexing is performed for each.

2 Research objective

Suppose we have simple and separate indices for pictures and sound of video information. Our target is to search for designated video object utilizing those simple indices.

By indexing, it becomes possible to search for a picture in which a person identified as Mr. X is reflected or to search for a picture in which Mr. X is talking (actually, this is a research theme in itself, but here we proceed with the research from the time point of its completion). Here, when we search for "a picture in which a person identified as Mr. X is reflected," Mr. X certainly appears over the picture, but there is no guarantee that Mr. X is actually playing a central role. Perhaps Mr. X is merely being reflected, while others are talking. In the same way, when we search for "a picture in which Mr. X is talking," a scene is certainly found where Mr. X is talking, but perhaps as a video, the result might be a picture with no relationship to Mr. X.

From the viewpoint of video processing or audio processing, the above may be sufficient because image or speech is involved. However, when we consider extracting coherent information from a video database, what we have is insufficient. In Section 3, we propose a "joint" operation for the purpose of extracting semantically coherent video objects, where "minced" simple indexes are used as clues.

Next, suppose simple indexing is performed for each of two independent video information, and consider how to search for video objects spread over these two groups. The example includes video of a satellite conferencing system. With this system, motion pictures and sounds are sent from two different base stations, and a conference is held in which base stations at various locations participate. The conference participants view a video 1 and a video 2 at the same time. In the figure, both video 1 and video 2 show the participants at two different locations conversing with each other.

If each video is recorded on separate media, for example, two video tapes, and digitized, both conference locations become logically separated, and the two participants who are talking appear completely separated in the other video. Assuming simple indexing on each, what method is adequate to grasp these two people actually talking? This is another example of extracting semantically coherent video objects (with simple indexes as clues). In Section 4, by using this example, we show a solution and demonstrate that the "joint" operation effective for this purpose too.

3 Operations on heterogeneous media data

3.1 Retrieving video objects

We discuss here how to utilize indices separately constructed on heterogeneous media data, such as motion pictures and audio, in order to retrieve meaningful video fragments. We call them "video objects." In the following discussions, we assume that simple indices on individual media, for example, indices on motion pictures representing which characters are on the screen and indices on audio representing which characters are talking, are given in advance.

Generally speaking, we cannot realize the retrieval of video objects as long as we individually use indices on each medium. Let us consider an example shown in Fig. 1. The figure depicts a video data indexed with time intervals. The x-axis represents the time line, and each time interval represents the index where some activities took place in each media, for example, a represents the time when Mr. X spoke, and b represents the time when Mr. X appeared. Mr.X's talk is represented with two time intervals, namely, a and c, because his voice was interrupted. Similarly, Mr.X's appearance is represented as three time intervals, namely, b, e, and e, because he disappeared from the screen due to the camera motion. Suppose we want to retrieve video objects using that index. We would like to treat b and d as a video object since his voice is continued even if he disappeared from the frame of the motion picture. To this end, we introduce "joint" operation in addition to conventional operations on time intervals such as intersection and union.

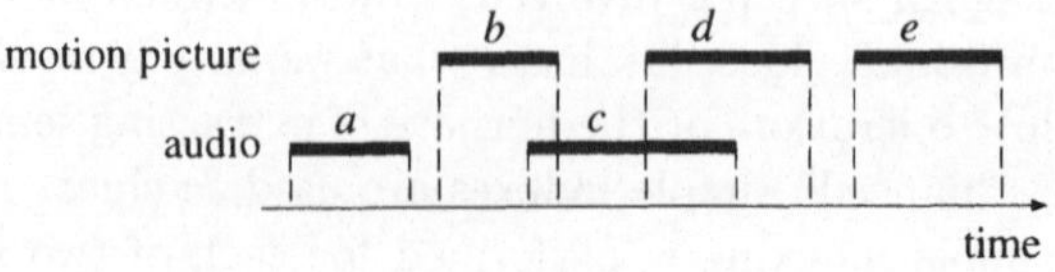

Figure 1: Simple indices

3.2 Operations on time intervals

In this subsection, we give formal definitions of time intervals and operations on them.

Definition 1 (time intervals) *Let a_s and a_e be time instants such that $a_s < a_e$. Time intervals are defined as (a_s, a_e).* □

In the next, we introduce three operations on time intervals, namely, union, intersection, and joint operations.

Definition 2 (union) *Let $A = (a_s, a_e)$ and $B = (b_s, b_e)$ be time intervals. Union ($\cup$) of A and B is defined as*

$$A \cup B = \begin{cases} (a_s, a_e), (b_s, b_e) \text{ if } a_e < b_s \text{ or } b_e < a_s \\ (min(a_s, b_s), max(a_e, b_e)) \text{ otherwise.} \end{cases}$$

□

Note that intersection operation returns two time intervals if its arguments are disjoint, that is, there are no overlapping interval between the arguments.

Definition 3 (intersection) *Let $A = (a_s, a_e)$ and $B = (b_s, b_e)$ be time intervals. Intersection ($\cap$) of A and B is defined as*

$$A \cap B = \begin{cases} \phi \text{ if } a_e < b_s \text{ or } b_e < a_s \\ (max(a_s, b_s), min(a_e, b_e)) \text{ otherwise.} \end{cases}$$

□

This operation returns a null-time interval, denoted by ϕ, if the arguments are disjoint.

Definition 4 (joint) *Let $A = (a_s, a_e)$ and $B = (b_s, b_e)$ be time intervals. Joint ($\oplus$) of A and B is defined as*

$$A \oplus B = \begin{cases} \phi \textit{ if } a_e < b_s \textit{ or } b_e < a_s \\ A \cup B \textit{ otherwise.} \end{cases}$$

□

The definition of joint operation differs from that of union operation in the case that the arguments do not have any overlapping interval, then it returns a null-time interval. We can naturally extend joint operation among more than three time intervals. Let $S = \{A_1, A_2, \ldots, A_n\}$ be a set of time intervals. Joint of S can be calculated by applying intersection operation to each overlapping pairs. The following is the detailed depiction of the algorithm.

Algorithm 1 (joint of time intervals)
(1) Sort all elements $A_i (1 \leq i \leq n)$ in S according to their start times.
(2) for $i = 1$ to n; do
    $A_i \oplus A_{i+1} \rightarrow comp$;
    if $comp = \phi$ then
        output A_i as an element of the result;
    else
        $S = S - A_i$;
        $comp \rightarrow A_{i+1}$;
    endif
  done.

4 Integration of media objects according to time axes

4.1 Motivation

In practical, there may exist some videos recording the same event in the real world. In such cases, it is useful if we can extract video fragments using information from all video data. Let us consider the following situation. Suppose we have two simultaneously recorded video data in which a satellite video conference is recorded, and they are taken at distant places. No one can hence appear on both of these videos at the same time. Indices individually constructed on those videos therefore have no relation with each other. For this reason, we cannot retrieve any video object that consists of both video data. Although we can make up an index according to both of them by hand, it should require large cost.

In order to overcome this problem, we propose a way to relate independent video data. As a result, we can process queries like below:

1. Find the video objects in which Mr. X talked with Mr. Y.

2. Find the video objects in which the person who talked with Mr. X appeared.

4.2 Classification of time axes

Here we discuss classification of time axes in video data, which is necessary to realize query retrieval of video objects. Iketani et al. classified time axes of video data into the following four categories, namely, "media time," "story-time," "record time," and "display time." Because the story time is intended to model time axes related to stories in cinemas and dramas, we redefine it as "semantic time" in order to cope with any kind of video. In addition, we newly introduce "logical time." As a result, we obtain the following five kinds:

1. **Media-time axis:** Video data usually reside in storage devices. The media-time axis corresponds to the order in which a video data is stored in its storage device. It also reflects its physical structure in the storage. Thus, it begins at 0 second and continues for the playback time of the stream. Note that we can establish the media-time axis of the video data even if it is stored in discontinuous devices, such as hard disks, as long as we can continuously access it in the direction of playback.

2. **Semantic-time axis:** The semantic-time axis represents virtual time axis created by the authors of a video, such as cameraman and director. In general, fictions including cinemas and dramas have complex temporal structure. Consequently, the passage of time in the video may dynamically change according to the scenario. The semantic-time axis is used to relate external information, such as places and events, to the video contents.

3. **Record-time axis:** The record-time axis of a video data represents the time when the video data was recorded. People who edit unfinished videos often use it, but most of the users who just see completed products do not.

4. **Display-time axis:** The display-time axis of a video data represents the time along which a video is shown. A user uses it to specify the time when the video is played. In this sense, the display-time axis is independent of the former three axes, and we can specify arbitrary time for it.

5. **Logical-time axis:** The logical-time axis of a video data includes correspondence between semantic-time axis and its external information. It is useful to model video data which have same semantic-time axes and different external information. Let us consider an example of satellite video conferencing discussed above in that there are two simultaneously recorded videos of video conference and sightseeing. Although these videos happened to have the same semantic-time axes, the contents are different. Consequently, it is important to relate the semantic time with its corresponding external information.

4.3 Classification of video data

We can define the following categories for the types of video data using the above mentioned time axes.

1. **Material type:** Video data in this category have the same sequences of time intervals on the media-, record-, semantics-, and logical-time axes. That is, they should be raw videos just recorded, are not edited, and consist of one scene.

2. **Relay type:** Video data in this category have almost the same sequences of time intervals in the record-, semantics-, and logical-time axes. A relay broadcast of satellite video conference is a typical example. For videos in this category, there may be gaps on record-time axis. However, There are no alternation in the sequences of time intervals on the media and record-time axes.

3. **Cinema type:** The media-, semantics-, and record-time axes of video data in this category are independent. Most of cinemas and dramas are categorized into this type.

4.4 Joint operation on logical-time axes

Utilization of logical-time axis enable us to retrieve video objects from several videos by interrelating separated video segments in terms of logical-time axes as described above. However, suppose we have several videos of a satellite video conferencing and a sightseeing in a video database. If there is no causal relationships between the conference and the sightseeing, then their logical-time axes do not overlap. As a consequence, we cannot apply joint operation.

In fact, it is a quite complicated problem how to represent logical-time axis. Conceptually, it is a natural way to represent logical-time axis as the combination of semantic-time axis and identifiers representing events which are related to the video, such as "conference" or "sightseeing." In databases, they can be represented as a set of time intervals on the semantic-time axis, the serial number of the video data with the name of the relation representing the identifier of the video data. In this way, however, we cannot express the video data only with time intervals.

In this chapter, we employ satellite video conference as an example for observation and experiments. For this reason, we do not need to consider logical-time axes, since we can consider semantic- and logical-time axes are identical in this type of video data, as described above.

Let us consider the queries stated in Section 4.1. We can process the first query by the following operation:

1. Extract time intervals from one video data's index that includes Mr.X's appearance and his voice, and apply joint operation on them. Perform the same operation to the other video data for Mr. Y.

2. Apply joint operation to the two sets of time intervals obtained by 1.
3. Repeat 1 and 2 by swapping Mr. X and Mr. Y.

It is important to notice that time intervals used in this operation must be on semantic-time axes, unless the result of this operation may lost its meaning. In addition, the result just represents the time when t and u shared their logical time. Consequently, it is not guaranteed that they had a conversation in that time. In other words, at least the time when they talked must be included in the result. The quality of the result depends on the accuracy of the indexes.

We can process the second query as follows:

1. Extract time intervals from one video data's index that includes Mr. X's appearance and voice, and apply joint operation on them.
2. For each time interval included in the result of 1, apply joint operation to all time intervals in the index of the other video data.
3. Apply 1 and 2 to another video stream.

As described before, the result of this operation is also approximation.

5 An overview of a video database system "Heijo"

5.1 System configuration

So far, we have implemented prototype video database systems named "Kasumi" and "Asuka" by constructing data models for logical and physical aspects of video data, respectively. In this study, we reconstruct the whole system, and implement a new prototype system "Heijo." Heijo is designed as a distributed system consisted of UNIX and Windows systems.

Fig. 2 shows the architecture of Heijo. The architecture of Heijo has the following features:

1. It is a distributed system composed of a UNIX server and Windows clients.
2. It employs a commercial DBMS Illustra with Web DataBlade as the database engine. For this reason, a user of Heijo can utilize Web browsers as the interface this system.
3. It supports MPEG-1 video format.

The functional features of Heijo is that it can retrieve semantically coherent video objects based on simple indices. Precisely, we have implemented some operations such as "joint" and "talk-with," a variation of "joint."

Each component of Heijo is connected with TCP/IP based network. The following subsections briefly explain the components.

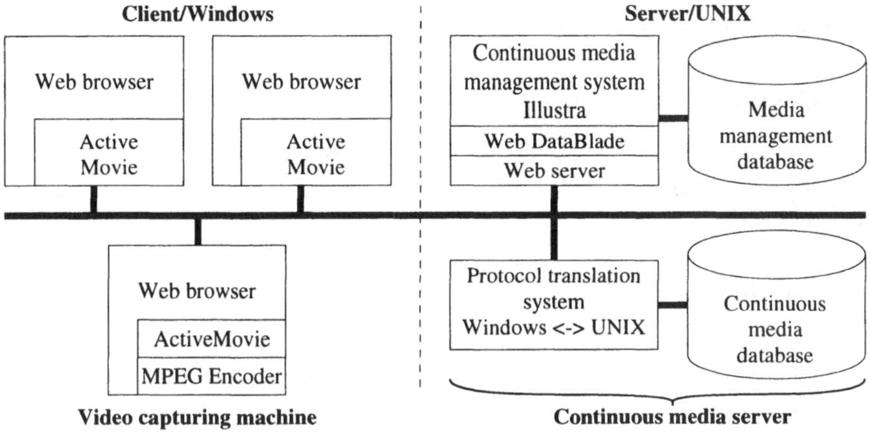

Figure 2: System configuration of Heijo

5.1.1 Clients

We employ Internet Explorer, a Web browser available on Windows 95/98, as the user interface of Heijo. Video data are played back using ActiveMovie, which is a plug-in program of Internet Explorer. Thus, users can perform miscellaneous special playback, such as "rewind" and "fast-forward," using the functionality provided by ActiveMovie in addition to ordinary playback operation like "play" and "stop." Note that the name of this product has been changed from "ActiveMovie" to "DirectShow" from Ver. 2.0. We, however, use "ActiveMovie" in the followings, because our system uses Ver. 1.0.

5.1.2 Video capturing machine

We use a client with MPEG encoder board as the video capturing machine. Users can capture video data, convert the data to a MPEG video, and create indexes on the movie.

5.1.3 Media management database

The media management database maintains indexes for each media data, and is constructed on the continuous media database (5.1.6) which plays an important role of servers in UNIX side.

5.1.4 Continuous media management system

The continuous media management system manages handles media data together with the continuous media server (5.1.5), and actually processes operations on temporal intervals and integration of media data on the same time-axis so that we can utilize video data in continuous media server.

5.1.5 Continuous media server

The continuous media server consists of the protocol translation system (5.1.7) and the continuous media database (5.1.6). It sends video data to the client if requests from clients arrive.

5.1.6 Continuous media database

The continuous media database stores video data itself. Actually, it is implemented on UNIX file system in the current system. Clients can access this database via the protocol translation system (5.1.7).

5.1.7 Protocol translation system

The protocol translation system translate video data of UNIX file system into those of Windows file system, and vice versa, so that clients running on Windows can access the continuous media database.

5.2 Capturing, storing, and querying video data

Table 1 shows the storage requirements for major video formats. For example, NTSC is a TV-quality format, and HDTV is a high quality TV format. HDTV requires 1TB to record video data of a little less than 2 hours. At present, MPEG-1, in which transmission rate is fixed, has widely been used, and MPEG-2, which has variable transmission rate mechanism, is about going to diffuse. However, 1TB storage space can provide only video data for 280 hours, even if we use MPEG-2 with 8Mbps in which we cannot recognize the decline in its quality. If we assume that the total length of one lecture is 20 hours, we can store 14 lectures.

Table 1: Storage requirements for misc. video formats

format	data rate	storage requirements	available time per 1TB
NTSC	100Mbps	45GB/h	22h/TB
HDTV	1.2Gbps	540GB/h	1.8h/TB
MPEG-1	1.5Mbps	675MB/h	1470h/TB
MPEG-2	8Mbps	3.6GB/h	280h/TB

Let us consider a video data of VGA size (640 × 480 dots). In order to record this video data, 27MB/sec for bandwidth and 97GB/hour for storage are required. Thus, 97TB is necessary for recording video data of 1000 hours.

Although magnetic disks have been getting higher density as technology in storage devices rapidly advances, we must depend on data compression for several or up to ten years in order to construct video databases.

Raw video data, that is to say video data before compression, can be considered as an ordered set of still images so called frames. We can therefore extract arbitrary frame from the them. On the other hand, frames in compressed videos, such as MPEG-1 and MPEG-2 data, are no longer independent, and we cannot simply extract them. The minimum unit in which we can extract arbitrary frame is called GOPs (group of pictures), and is about 1/2 sec fragment in case of MPEG-1. We have constructed a model for compressed videos by taking the above mentioned characteristics of MPEG-1 into account, and have implemented a video database system which can extract arbitrary frame based on this model. We also have employed MPEG-1 as the video format for Heijo. But the operation completely depends on the facility of ActiveMovie, and thus the minimum unit in which we can direct is one second.

Users of Heijo can capture a video, and compress it as MPEG-1 data in real time using a Web browser (Internet Explorer).

In the current system, we must create all indexes by hand. We are planning to develop automatic generation of indexes in the next step. However, we have implemented a system to facilitate indexing.

6 Conclusions

One of the most important challenge of video databases is how to recognize meaningful segments of video data. In this chapter, we have discussed the way to capture meaningful video data using simple indices. To this end, we have proposed the "joint" operation, and two methods for utilizing this operation. Finally, we have reported our prototype video database system "Heijo."

We have just examined our system by an example of satellite video conference, which is relatively simple broadcast type video data. For this reason, it is necessary for us to evaluate our system by more complex examples. Moreover, future work includes: (1) evaluation of expressive power of our operations, (2) stricter formalization and modeling, and (3) establishment of operations necessary and sufficient for acquisition of video objects.

Bibliography

1) J. F. Allen. Maintaining knowledge about temporal intervals. *Commun. ACM*, 26(11):832–843, Nov. 1983.
2) T. Amagasa, M. Aritsugi, Y. Kanamori, and Y. Masunaga. Interval-based modeling for temporal representation and operations. *IEICE Trans. on Info. & Syst.*, E81–D(1):47–55, Jan. 1998.
3) T. Kanade, S. Satoh, and Y. Nakamura. Informedia: CMU digital video library project. *IPSJ Magazine*, 37(9):841–847, 1996. (in Japanese).

4) T. Kanade, S. Satoh, and Y. Nakamura. Accessing video contents: Cooperative approach between image and natural language processing. In *Proc. Int. Symposium on Digital Libraries 1997 (ISDL '97)*, pages 143–150, 1997.
5) P. Ladkin. Time representation: A taxonomy of interval relations. In *Proc. the National conference on Artificial Intelligence*, pages 360–366, 1986.
6) B. Leban, D. McDonald, and D. Forster. A representation for collections of temporal intervals. In *Proc. AAAI-1986 5th Int. Conf. on Artificial Intelligence*, pages 367–371, 1986.
7) Y. Masunaga. A unified approach to rrepresentation, synchronization and storage of temporal multimedia objects based on time interval logic. In *Proc. 5th Int. Conf. on Database Systems for Advanced Applications (DASFAA '97)*, pages 353–362, Apr. 1997.
8) E. Oomoto and K. Tanaka. OVID: Design and implementation of a video-object database system. *IEEE Trans. on Knowledge and Data Eng.*, 5(4):629–643, Aug. 1993.
9) R. Weiss, A. Duda, and D. K. Gifford. Content-based access to algebraic video. In *Proc. Int. Conf. on Multimedia Computing and Systems*, pages 140–151, 1994.
10) R. Weiss, A. Duda, and D. K. Gifford. Composition and search with a video algebra. *IEEE Multimedia*, 2(1):12–25, 1995.

7

Mediator-based Modeling of Real World's Objects and their Motions

Hiroshi Arisawa
Takashi Tomii

Department of Electrical and Computer Engineering
Faculty of Engineering
Yokohama National University

ABSTRACT

The author's group has beed developing next-generation multimedia database called Real World Database(RWDB), in which not only image and audio data but also object structure, behavior, temporal event and all spatial occasions are stored uniformly. In this paper, we will concentrate ourselves into the modeling methodology of such spatio-temporal objects and their motions. For this purpose, we introduce "Mediator" concept to describe object's and motion's characteristics with small amout of data. The mediator is an approximated model based on the meaning of the objects, with semiautomatic fitting on the on the sampled data from real world. This way it becomes possible to supply general information about a certain type of objects such as human bodies and integrate the individual parts data into a meaningful model representing each instance. In this presentation, two more medators are discussed: Structure Mediator and Motion Mediator. These Mediator data can be also detected semi-automatically from real world. Finally, we will show the concrete database schema to represent relationships between three types of Mediators and real world data.

1 Introduction

The author's group has been developing next-generation multimedia database called Real World Database (RWDB), in which not only image and audio data but also object structure, behavior, temporal event and all spatial ocations are stored uniformly. In this paper, will be focusing on the modeling methodology of such spatio-temporal objects. In order to realize such "real world" database system, we must consider at least four components as follows.

- Real World Capturer (RWC)
- Real World Modeler/Analyzer (RWM)

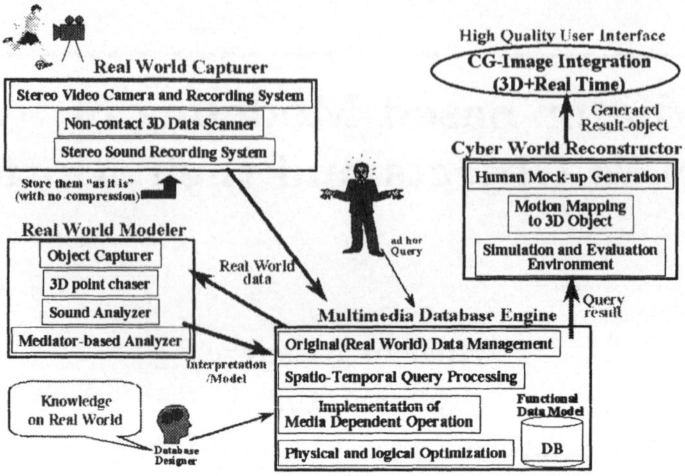

Figure 1: Conceptual architecture of RWDB

- Multi-Media DataBase Engine (MMDBE)
- Cyber World Reconstructor for retrieval (CWR)

Among them, the RWM has an essential role to recognize the object and the motion from captured data and to extract "abstract"(or index) data for database retrieval. For this purpose, we introduce "Mediator" to describe object's and motion's characteristics with small amount of data. The mediator is an approximated model based on the meaning of the objects, with semi-automatic fitting on the sampled data from real world. For example, as to figure of human head, we use a cylinder with 10 layer, polygonal column[1)]. Model of a human head has the additional information that the third layer from top is the height of eyes. Also, it is also possible to supply position information that in the parameter of these sections which vertex represents the left and right eye or the mouth. This way it becomes possible to supply general information about a certain type of parts, and integrate the individual parts data into a meaningful model representing each instance.

Additionally, this paper describes the RWDB concept, and is focusing on Modeling aspect of real objects. The (shape) Mediator is proposed as an optimum solution. Also, "event" concept is discussed from the viewpoint of "location analysis" of objects. In this paper, two more mediators are discussed: Structure Mediator and Motion Mediator. These Mediator data can be also detected semi-automatically from real world. Finally, we will show the concrete database schema to represent relationships between three types of Mediators and real world data.

2 Real World Database

The objective of the RWDB is to provide a total environment of capturing, modeling and storing physical or logical objects in the real world. Everything in the real world could be modelled through it and any type of data could be accumulated. For this purpose, RWDB must involve at least 4 components listed below. Conceptual architecture of RWDB is shown in Figure.1.

- Real World Capturer (RWC)

  The objective of RWC is to capture the external form of objects in the real world. There exists various types of 3D or spatial information depending on capturing devices. The simplest one is a set of video cameras. We can get a sequence of frames from, for example, a left-eye camera and a right one simultaneously. Recent technology enables us to get the full-color and high quality image with no compression of data. However, for the frame sequences, it is very hard to extract a complete surface model or motion (kinematics) model of each object. On the other hand, several "Motion Capturing Systems" are commercially available but those system can trace only a small number of marking points and the most part of original information (surface and texture) are lossed.

  Another type of input device is "3D Scanner" by which we can get a perfect surface (polygon) model for static objects. The practical solution is to get above two kind of informations from the real world and to combine two models into one in the database level. This idea is summerized in Figure.2.

- Real World Modeler (RWM)

  RWM is a set of tools each of which analyzes original frame images and generate a new information. For example, the Outline Chaser[3)] catches the outline of an object in a certain video frame, and then trace the outline in preceding and successive frames. On the other hand, the Point Tracer detects stereo pairs and the range values for (specified) points from a couple of (Left and Right) frame images, and makes a rough sketch of 3D objects in the real world. Many algorithms are investigated and evaluated for range image generation in Image Processing area [4)]. All the results of analysis are stored into database preserving the correspondences to the original images.

- Multimedia Database Engine (MMDBE)

  MMDBE is a database which treats a variety of data types such as full texts, graphic drawing, bitmap images and image sequences. The features of such data are quite different from the conventional DBMS's ones, because some of them are continuous and might occupy much more space than traditional data types. As to data model, in order to integrate all types of data, introduction of simple primitives to describe real world entities and the associations between them are essential. Moreover, the query language of multimedia data handling

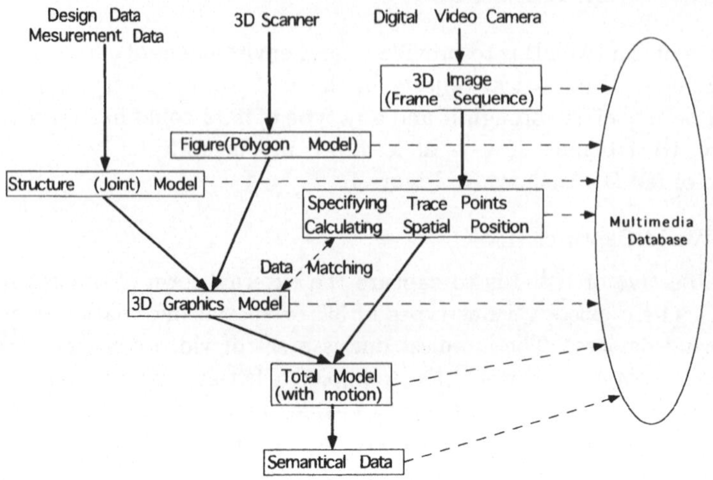

Figure 2: Integration of various types of 3D Data

must involve various types of media-dependent operations for retrieving and displaying. Especially, in RWDB, the result of a query creates a new "Cyberspace". For example, for a "Work Factory Database" which includes unit works of human workers, RWM might extract a worker's motion in each work. A query may retrieve a certain unit work and project it to another human worker. The author proposed a total data model to describe the 2D or 3D data, and also presented query language MMQL for flexible retrieval [4)].

- Cyber World Reconstructor (CWR)

  As discussed above, the result of database consists of various types of data such as frame sequence and 3D graphics data. In order to visualize the query result, RWDB should provide a "player" of result world. CWR is, in this sense, an integrated system of 3D computer graphics and 3D video presentation.

  Unfortunately, the modeling method of objects in the field of 3D graphics and VR systems are quite different from the DB approach because the former is focusing on natural and smooth presentation of surfaces of objects and their motions, whereas the latter makes deep considerations on semantic aspects os these objects.

3 Modeling of The Human Body and its Motions

Based on the above RWDB concept, we concentrate ourselves into modeling on human bodies and motions. We offered Human Working Model for storing Human motion data and reconstructing Human motion in CG Simulation.

3.1 Human Body and Motion Modeling

In order to store and retrieve Human motion data, we need an integrated datamodel which can describe all kind of objects and materials about the human body, for use in all applications. Creating realistic model of the human body requires considerable amount of data because of object's complexity. Therefore from ergonomical point of view we must re-modeling the human body with small number of primitives. We call it "Simplified Human Body", which involves, for example, simplified head, arms, body and legs, connected by small number of joints each other.

There are two phase in modeling of the human body and its motions.

3.1.1 Human Static Model

1. Polygon Model

   Human body's figure can be described by using Polygon and Texture Models. An example of polygon model alone is shown in Figure.3, and a complete texture image is shown in Figure.4.

Figure 3: Polygon Model of Head

2. Structure Model

   Human body consists of more than 200 bones, which are connected so that the body can move using the power of the muscles [6)7)]. But mechanism is too complex to model it completely. We focus on movable joints and select 24 of them in order to create a simplified model of the human body, Human Skeleton Model. Each child component of Human Skeleton Model has its own coordinate system, the origin of which is the connecting point (i.e. joint) to the parent component.

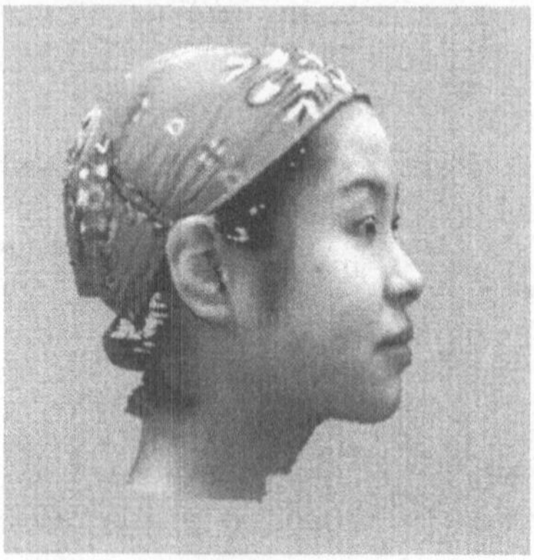

Figure 4: Texture Image of Head

An example of Human Skeleton Model and connections between parts are shown in Figure.5.

Table.1 shows an example of such parent-child relationships using a traditional style of Relational Database. For each child component, the parent component, the position (x, y, z) of joint and initial angles $(\theta_x, \theta_y, \theta_z)$ to the parent are saved.

Actual movements of the joints have strong connection with the muscles. In our research additional simplification of the considerations is achieved by considering only the range of motion for the selected 24 joints. Each joint has the Degree of Freedom less than three, rotation or translation. In Figure.5, the joints expressed by cycle can rotate. We can do this because it would be possible to compare the employee's work with the machine's work, especially regarding to their abilities to perform certain actions, if we regard the human body as a "rigid".

3.1.2 Human Dynamic Model

In Dynamic Model, Human Motions are described. For a certain employee's work, the action of each component of human body should be traced and modelled under the restriction of inter-joint structure. This Dynamic Model is defined on the Human Skeleton Model discussed above. Each joint move along with time. We can also use Relational Database to express the joint movement as a successive relative angles every child component. An example of movement of Tab.1's object following to Tab.2.

Table 1: Tabular Representation of Object Structure

Name	Parent	Relative Position(mm)			Initial Angle(degrees)		
		x	y	z	θ_x	θ_y	θ_z
Base	–	0.00	0.00	0.00	0.00	0.00	0.00
LOIN	Base	0.00	0.00	881.1	0.00	0.00	0.00
TORSO	LOIN	0.00	0.00	15.00	0.00	0.00	0.00
CHEST	TORSO	0.00	0.00	200.00	0.00	0.00	0.00
NECK	CHEST	0.00	0.00	157.20	0.00	0.00	0.00
HEAD	NECK	0.00	0.00	65.50	0.00	0.00	0.00
L_SHOULDER	CHEST	0.00	65.00	207.20	0.00	0.00	0.00
L_UPARM	L_SHOULDER	0.00	1044.80	0.00	-180.00	0.00	0.00
L_FOREARM	L_UPARM	0.00	0.00	331.50	0.00	0.00	0.00
L_HAND	L_FOREARM	0.00	0.00	269.20	0.00	0.00	0.00
R_SHOULDER	CHEST	0.00	-65.00	207.20	0.00	0.00	0.00
R_UPARM	R_SHOULDER	0.00	1044.80	0.00	-180.00	0.00	0.00
R_FOREARM	R_UPARM	0.00	0.00	331.50	0.00	0.00	0.00
R_HAND	R_FOREARM	0.00	0.00	269.20	0.00	0.00	0.00
L_THIGH	LOIN	0.00	81.00	0.00	180.00	0.00	0.00
L_LEG	L_THIGH	0.00	0.00	377.10	0.00	0.00	0.00
L_FOOT	L_LEG	0.00	0.00	43.34	0.00	0.00	-30.00
R_THIGH	LOIN	0.00	-81.00	0.00	180.00	0.00	0.00
R_LEG	R_THIGH	0.00	0.00	377.10	0.00	0.00	0.00
R_FOOT	R_LEG	0.00	0.00	43.34	0.00	0.00	30.00

Table 2: An example of Joint Value of Work

Time	Joint1 (deg)	Joint2 (deg)	Joint3 (deg)	Joint4 (deg)	Joint5 (deg)	... (deg)
0.00	4.286	3.249	0.000	32.57	0.00	...
0:10	4.286	3.249	0.000	32.57	0.00	...
0:20	4.286	3.249	0.000	32.57	0.00	...
0:30	4.286	3.249	0.000	32.57	0.00	...
0:40	5.265	4.228	0.000	32.57	0.00	...
0:50	5.286	4.249	0.000	32.57	0.00	...
0:60	5.332	4.295	0.000	32.57	0.00	...
1:00	5.846	4.809	0.000	32.57	0.00	...
1:10	6.286	5.249	0.000	32.57	0.00	...
...	...	...	...	...	...	...

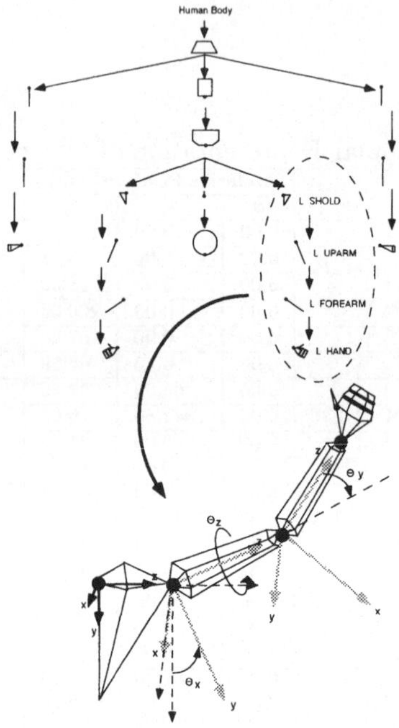

Figure 5: Structure Model

3.2 Event Modeling

When the Motions of Human Model are described, two aspects have to be discussed. The first one,Human Model's parts and joints are moving, is discussed above. The other, characteristic of namely the way motions of their Semantics, for example, driving in a nail, lifting an object, writing characters, etc. We call event everything that occurs, any significant activity during the production process.

When considering events, depending on the user's objectives, different levels of abstraction are applied. For instance, we can consider as an event simply the screwing of a screw or the making of a whole car. In the field of Industrial Engineering such separation is strictly regulated. The same production process is regarded at four different levels — production line, processes, elementary operations, and elementary movements, and there are hierarchical relations between them. An example is given, in Figure.6. Thus, events should be defined in a way which reflects such multilevel organization. It should be added this is the natural way people think.

Apparently, some hierarchical structure is suitable for expressing event's meaning. Each event should be considered as consisting of sub–events, which themselves are events, too. A tree seems to be a good choice but here are also some other facts to consider. What we mean is that very often there are identical or very similar parts between several different events. Their number increases is inverse proportion to the level of consideration, i.e. the lower is the level, the more are the repeating sub–events. Keeping the number of redundancies as small as possible is a major characteristic of the database systems, so avoiding such a big number of repetitions of the same event is an important criterion. One possible solution is to allow each event to be regarded as a part of several higher level events. Therefore, the appropriate structure which fits to the above conditions appears to be a hierarchical network. In Figure.7 is shown a concrete example. Choosing network as a basic structure for describing events imposes some special requirements on the data model and database query language. The data model should provide simple and natural mapping of the hierarchical network into the data model primitives.

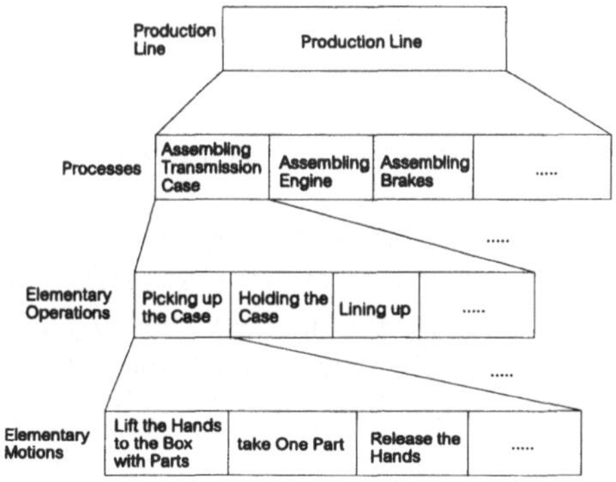

Figure 6: An Example of Event Model

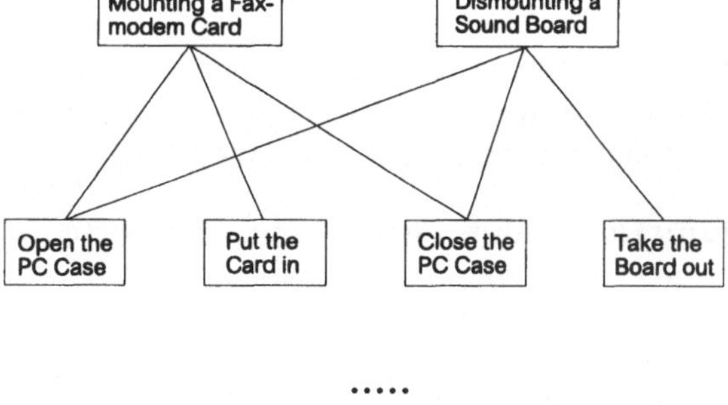

Figure 7: Event Tree

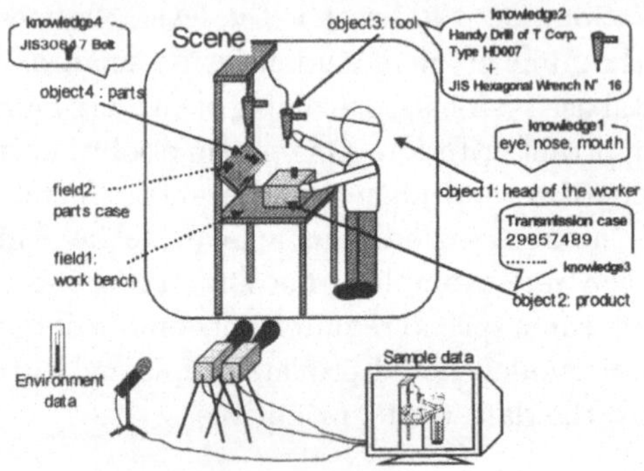

Figure 8: Concept of a Scene

4 The structural elements of a scene

In order to transform the information of the actual world to a database, first it is necessary to think what kind of spatio-temporal information should be observed and what kind of data can be implemented with. Here we will call the range of the world that serves as the unit for database transformation a **scene**. As Figure 8 shows, the scene information is divided into the following levels.

- **Information about the scene** Fundamental information about the world to be transformed, including spatial and temporal range, the standard position of the coordinate axes, etc.

- **Information about the scene construction** Static information about the place where the objects that take part in the scene move, e.g. workbench, fields, etc.

- **Information about the major objects taking part in the scene** Dynamic information concerning the shapes and movements of the objects that take part in the scene, e.g. workers, robots, tools, balls, etc.

- **General information about a given type of objects** General knowledge data about the objects, leaving out the individual details. For example in case of humans the general place where the nose is on the head, etc.

- **Sample data taken from the world outside the scene** Sample data taken by video cameras, three-dimensional scanners, manometers, thermometers and other kinds of measuring units. Most certainly, the sample data itself is not supplied by any kind of information about the movement or meaning of the objects that take part in the scene.

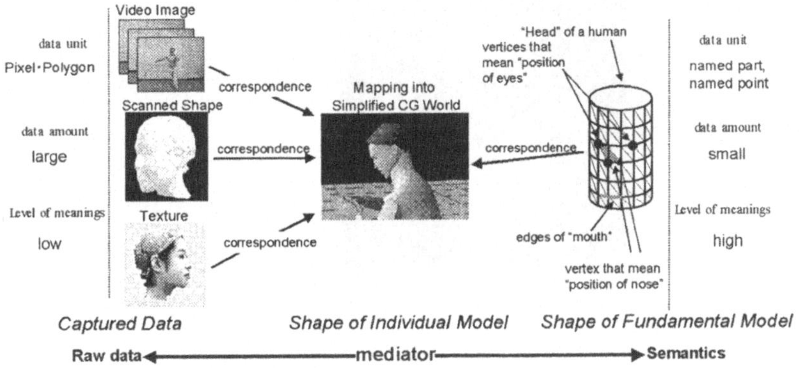

Figure 9: Acquisition of the Scene Data

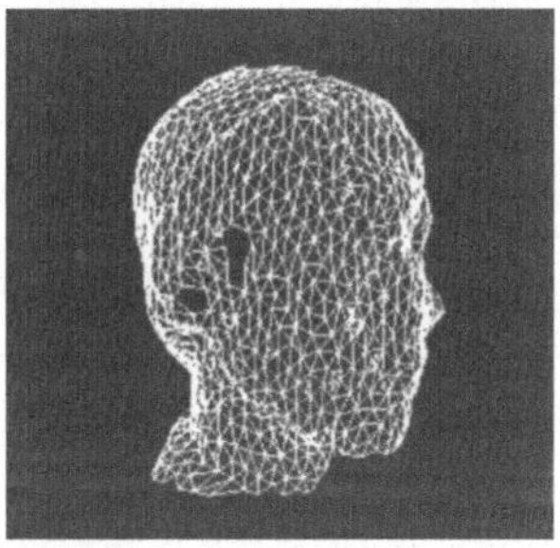

Figure 10: Scanned shape

In order to transform any real world space-time information data into a database, it is necessary to model and store all the information of all the above levels.

5 Constructing scene databases

5.1 Modeling technique

As Figure 9 shows it, in our present research we introduce an approximated model based on the meaning of the objects, with semi-automatical fitting of the sampled data and with so-called mediators that bridge the gap between the sample data and its meaning. These mediators represent the individual objects of the scene with relatively good fidelity yet few polygons. We tried to construct a scene database that maps these into virtual CG space [13].

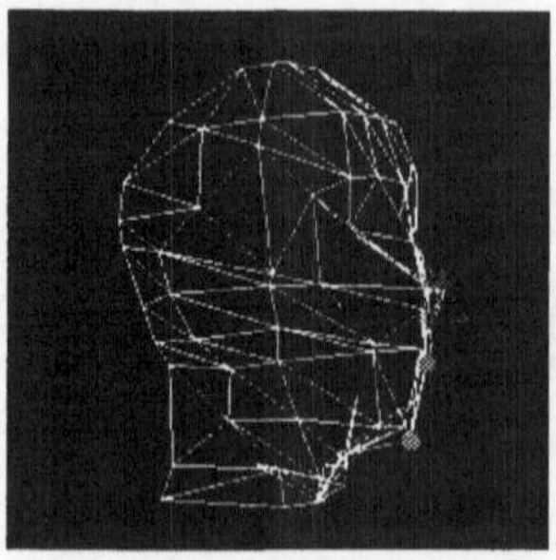

Figure 11: Generated Shape of Individual Model

Many levels and types can be thought of the data taken from the real world. In our present research we made our goal to store the shape and the position of the objects from the target world and their changes in time. Particularly, in the case of analysing human motions and the environment or devices that may affect those motions [15). It becomes impossible to capture the correct movement information directly. From the viewpoint of the acquisition of the work and movement information, we would like to presuppose here that data is taken without any special equipment. For that reason here we acquired data from the outer world into the scene only from three-dimensional pictures taken by multiple cameras, and the correct shapes and textures of the objects we determined by using a non-contacting type, three-dimensional scanner.

Most certainly, in this kind of sampled data there is not yet any "meaning information", such as in the video frame which pixels are actually the objects or in case of shape data, which polygon group is representing the nose of a human being. Only the humans can and have to specify and store "meaning" the human knowledge or interpretation of the objects together with the other kinds of information about the objects. However, as the pixel and polygon data is fairly excessive in amount, it is not really feasible for humans to specify "meaning" manually, place by place, in detail.

Here we propose a three-level concept about the shapes of the objects that supplies a semi-automatic process of "adding meaning" together with storing data in the database and data retrieval.

- **Shape of Fundamental Model** We will call "Shape of Fundamental Model" the rough shapes that express general knowledge (with minimal internal details) with relatively little polygon data about objects e.g. "human parts" (an eye, a nose, a mouth or an ear) or a hammer that consists of a metal part and a handle. For example, in our present research we use a "10-layer, dodecagonal column" as Shape of Fundamental Model of a human head, which has the additional information that the third layer from below is the position of the

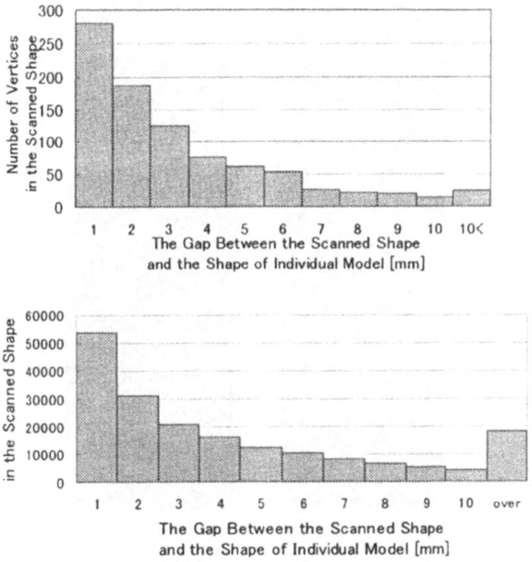

Figure 12: Histogram of the Gap Between the Scanned Shape and the Shape of Individual Model

lower jaw and the second layer from above is the height of eyes. Also, it is also possible to supply position information that in the perimeter of these sections which vertex represents the left and right eye or the mouth. (Figure 9). This way it becomes possible to supply general information about a certain type of parts, and integrate the individual parts data into a meaningful Shape of Fundamental Model.

- **Scanned Shape** It is possible to acquire very delicate but static shape data consisting of several hundred thousands of polygons about the individual objects with a three-dimensional scanner. We call this the "scanned shape". It is possible to reduce the number of polygons of the scanned shape and apply textures on them and thus create a graphics model of arbitrary quality, but the scanned shape being one kind of the sampled data, it does not have any information about meaning. Also, as the individual sampled data is not unified, applying "meaning" to it becomes relatively computation-intensive, thus it is not quite viable. Therefore, if meaning-based data representation and manipulation is required, then it is necessary to create the so-called Shape of Individual Model we detail below.

- **Shape of Individual Model** On one hand, the Shape of Fundamental Model reflects "meaning", while on the other hand it represents the scanned shape

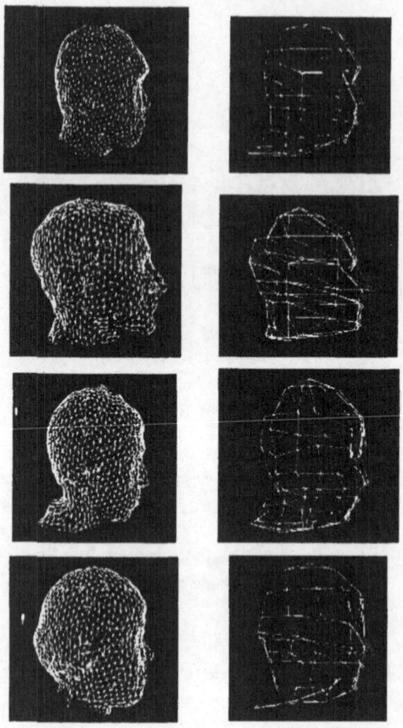

Figure 13: Examples of Scanned Shapes and Shapes of Individual Models

with approximated polygon data. In other words, the Shape of Individual Model is a mediator that provides a link between the data sampled directly from the real world and the model for database construction. The Shape of Individual Model has the shape of the individual objects and because it is created with relatively few polygons, a virtual CG space can be constructed with the Shapes of Individual Model, in other words, a scene database. As it is possible to measure the "unity" among the individual objects, it is particularly useful and effective in data retrieval and comparative evaluations. As a result, the "meaning" of each polygon is common in the all shapes of Indivisual Objects. For example, No.32 polygon always means the position of the left eye and No.66 polygon means the top of nose and so on. This is very important in the retrieving by using graphic information.

5.2 The acquisition and evaluation of the Shape of Individual Model

Based on the aforementioned theory, we created an experimental Shape of Individual Model acquisition system and evaluated it.

1. Acquisition of the scanned shape

To acquire the scanned shape data, we used a non-contact type three-dimensional scanner made by Cyberware (COLOR 3D DIGITIZER 3030RGB/PS/LN) and scanned in a human head. With this three-dimensional scanner, we scanned in a cylindrical range of a height of 30 cm and a diameter of 55 cm. It projects a perpendicular red slit laser beam on the target object and it determines the shape of the target object by moving the laser all the way around and composing the picture from the observed laser beam from different angles. The raw data we acquire this way consists of 230,400 vertices and about 350,000 triangles which we unify into polygons by geometrical algorithms and thus acqure the frontal shape. As a result, we get the polygon data like shown on Figure.10. This shape has 888 vertices, and consists of 1814 triangles.

2. Construction of the Shape of Fundamental Model

   For a Shape of Fundamental Model of the human head, as it can be seen on Figure 9 we used a very simple dodecagonal column that has 10 layers , and we provided the following "meaning information" for the layers: the lowest layer is the lowest part of the head (i.e. the neck), Layer 3 is the beginning of the jaw, Layer 4 is the layer of the mouth, Layer 5 is the bottom of the nose, Layer 6 is the tip of the nose, Layer 7 is the layer of eyes, and the top layer is the top of the head. We supposed that Layer 2, 8, 9 and 10 are divided equally with those layers below and above them. Because there are vertices on the top and the bottom of the model respectively, this shape has 122 vertices and consists of 240 triangles.

3. Construction of the Shape of Individual Model

   Here first we manually adjusted the layers of the Shape of Fundamental Model with the corresponding parts of the scanned shape using necessary utilities. Then, in order to have these parts correspond to the parts of the scanned shape, we transformed the shape of the sections of the Shape of Fundamental Model. After the transformation, the process of changing the section shapes to polygons is automatical. As a result of this, we obtained the Shape of Individual Model as it can be seen on Figure 11. This shape has exactly the same number of vertices and polygons as the Shape of Fundamental Model. In comparison with Figure 10, the model reflects the original shape visually.

In order to evaluate to which degree the Shape of Individual Model we obtained this way reflects the scanned shape, for each vertex of the scanned shape we computed the error, i.e. the distance from the closest polygon of the Shape of Individual Model. The result of this can be seen in a histogram in Figure 12. The histogram shows that for most vertices (about 80%) of the Scanned Shape the error is less than 5mm. In case of a human head, such an error is enough for spatio-temporal querying and allows satisfactory visualization. More examples of Scanned Shape and Shape of Individual Model are shown in Figure 13. In all the cases, more than 80% of the vertices were also within the allowed mergin of error.

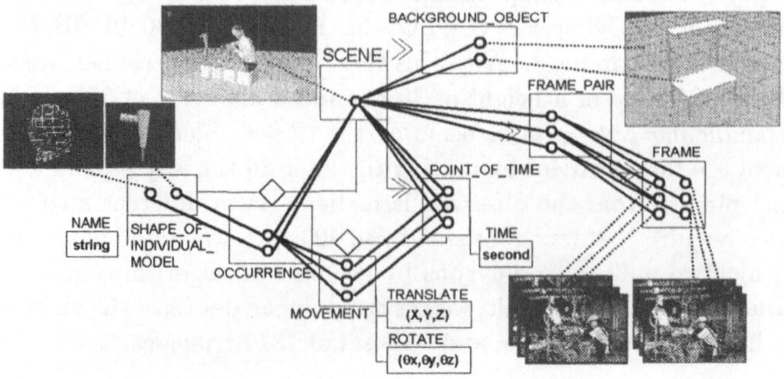

Figure 14: Schema and an Instance of the Scene Database

The Shape of Individual Model can be effectively used as a mediator but it is not feasible to create a mediator for every single object in the scene. However, the role of the mediator proposed in this paper is to provide index for the original data (shape data and pictures) when later querying the objects in the database. Thus, for indexing by information not expressed by the mediators, another indexing method is necessary. In our research, because with the original data the pictures themselves are stored, too, for those objects without a mediator though it imposes heavy computation costs at query time it is possible to use the pictures themselves to construct a cyberworld, and at the same time realize the same degree of reality.

6 Design and evaluation of the scene database

In this section we describe the database design based on the Shape of Individual Model and the movement information in the virtual space.

Figure 14 shows an example of the Shape of Individual Model mentioned in the previous section and how its movement data (in time) is expressed in the database. In the aforementioned scene database, the information (Shape of Fundamental Model, Shape of Individual Model) about the objects participating in the scene, the information about the location and the original sampled data (scanned shapes and three-dimensional pictures) and other primitives need to have a structure in which they are mutually and equally referring to each other. For this reason, it is convenient to use a data model having a flat structure instead of special, hierarchical structure. The relational model does have a flat structure and is capable of expressing two-way relations, but it is incapable to store "order" expressively and it is very difficult to use special operators graphic operators and picture operators, etc. that are necessary to manipulate the scene database. Also, the object-oriented model is based on a hierarchical structure, so it is suitable for handling graphics and pictures but at it is necessary to devise some way to express interdirectional references or other kind of "flat structure" information between the objects, it is hard to say that the object is always appropriate.

The goal of the present paper was to realize a way of core data modelling based on mediators that correspond to both the "meaning" and the actual data and to evaluate its feasibility. Therefore in this paper, to make it a test case of scene database design, with a design technique as much general-purpose as possible, and aiming to realize expression from the viewpoint of flat structure, we tried to express information based on an entity-based data model.

The schema database shown on Figure 14 is described using the AIS (Associative Information Structure) model[16]. The AIS model based on the ER model[17] is an entity-based data model; it expresses information using entities that represent the objects and phenomena of the real world and associations that express their relations. We already proposed a general data manipulation language for the AIS model, the MMQL [18], which can be used for database design from both the direction of data description and data manipulation. In Figure 14 the rectangles are expressing entity types and the small circles O represent entities of the given type. The relationships between the types can be direct (expressed by lozenges) multi-valued attributes (double arrow) , etc., and a solid line between entities shows correspondence of the instances.

The information expressed by the schema in Figure 14 is the following.

1. As information about the acquired scene (SCENE), we store the objects that make the background of the scene (BACKGROUND_OBJECT), the points of time in the scene (POINT_OF_TIME and the sampled data, including the pictures of the scene taken by a stereo camera pair (FRAME_PAIR, FRAME), etc.

2. We store the acquired Shape of Individual Model as an entity (SHAPE_OF _INDIVIDUAL_MODEL), with its polygon information.

3. We express the movement of the objects participating in the scene (MOVEMENT) by using entities that express the occurrence of the Shape of Individual Model of those objects in the scene (OCCURRENCE) at given points of time, and as a posture in these points of time, we store these with their coordinates and rotation angles (TRANSLATE, ROTATE).

In other words, the movement information of the objects in the scene is expressed as a series of position and rotation data of the Shape of Individual Model in time, and thus we can construct a scene database by mapping the scene into virtual CG space. As in this database the information about background objects building up the scene is expressed by a framework that unifies the sampled data (video pictures taken with a stereo camera pair, etc.), it is possible to realize several types of answers to the different questions, e.g. a video picture as a query result. As described above, we managed to construct a scene database that uses virtual CG space.

7 Motion Mediator

in section 5 we have introduced "mediator" to describe the shape of parts (ex. human head) of objects in abstract and uniform way.

Now, we consider to extend this idea to structural description and motions. Those are defined as follows.

- Structure mediator

  We can assume most of objects consist of several "solid" parts and "joints" connecting parts each other. For example, a human body consists of head, neck, shoulder, chest, right upper arm, left upper arm, and so on, and there are specific joints between those parts with the specific DOF(Degree of freedom). Structure Mediator is defined as joint values – range of motion and other constraint between parts – for each person. Structure Mediator can be regarded as an extension of shape (figure) mediator which can describe not only "simple" "part" objects but "complex" object like human body. This idea is shown in Figure.15.

- Motion Mediator

  Motion Mediator is defined as a sequence of intermediate postures within a motion which is done by a specific person. The intermediate posture must be preliminary assigned to each typical motion.

  For example,

  Unit Walking:

  I:Starting Motion

  II: Motion

  II–1 Right toe up

  II–2 Right toe at ground

  II–3 Left toe up

  II–4 Left toe at ground

  II–5 II–1

  III: Ending Motion

  are pre-assigned. (see Figure.16)

  The above intermediate postures can be detected by analyzing video images easily. Similar to the shape mediator, Motion mediator can represent the characteristics of a specific motion of a specific person, with a small amount of data. Therefore, from a set of motion mediators, the database user can calculate the average of the maximum height of right knee in a walking motion. The authors group is now developing a formal description of generic motion and a motion evaluation system in the framework of RWDB.

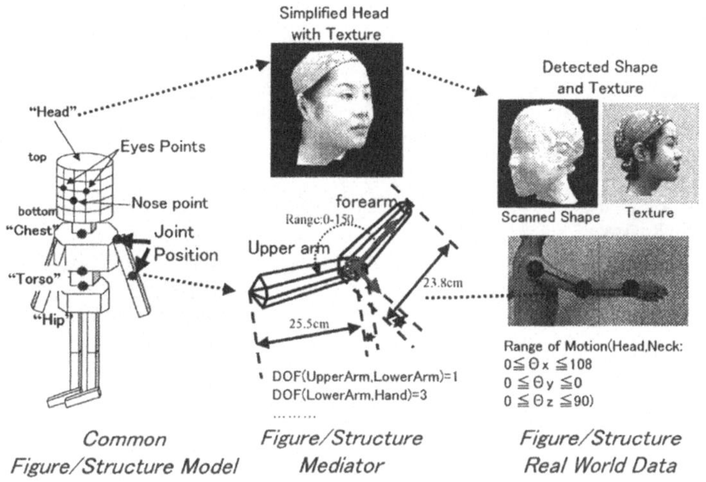

Figure 15: Common Model, Mediator and Real World Data about a figure and structure

8 Conclusion

In this paper we described a modeling techique for the RWDB based on the actual data and its design method. As for constructing the RWDB, first we divide the database by concepts called scenes. The database is constructed by first acquiring the raw data, video pictures and polygon information, then the scene information taken from this data is mapped into virtual space using the Shape of Individual Model. The Shape of Individual Model is created by a model that refers to the meaning of the actual data and therefore it can play the role of a mediator between the acquired data from real world and its meaning. Also, using the Shape of Individual Model as a mediator, we obtained the following results:

- We obtained an approximated shape with ample visuality
- We could perform query processing with ample precision
- We can perform aggregate functions based on the unified meaning
- The data acquisition from the picture can be performed by high precision and a low amount of computation

As a result, with this technique the RWDB is constructed according to a cyberspace using CG, the query processing and the spatial operations are based on the polygon objects, too. This way it becomes possible to compute the spatial relation

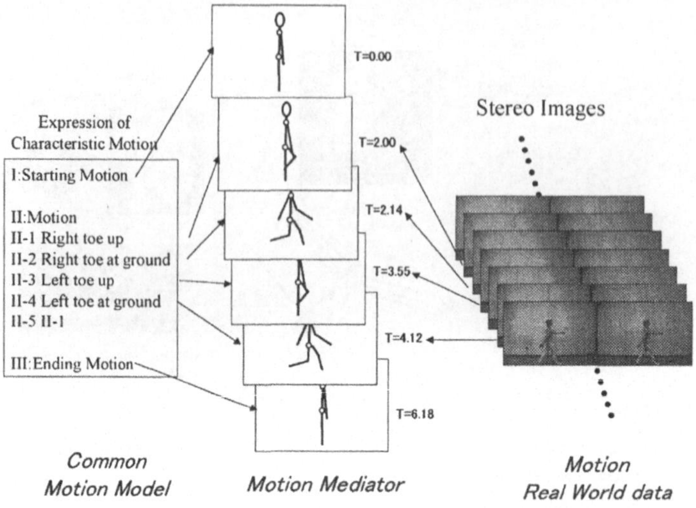

Figure 16: Common Model, Mediator and Real World Data about a motion

of the objects with fairly good precision and to use this cyberspace as the result of the queries.

At present we are developing a system that is capable of high-precision depth abstraction from the picture and posture abstraction of the objects using the obtained Shape of Individual Model. In this system it becomes possible to verify the evaluation of the mediators. The construction of a total system based on the RWDB using mediators, the design of time-dependant query processing, the implement of the RWDB and the evaluation of the spatio-temporal queries remain problems of the future.

Bibliography

1) Takashi Tomii, Szabolcs Varga, Sayaka Imai and Hiroshi Arisawa: "Design of Video Scene Databases with Mapping to Virtual CG Space", Proceedings of the 1999 IEEE International Conference on Multimedia Computing & Systems ICMCS'99, pp.741-746, Florence, Italy, 7-11 June, 1999.

2) Hiroshi Arisawa, Sayaka Imai : "Working Simulation based on Info-Ergonomics and Multimedia Database Concept Design" 1996 Japan–U.S.A. Symposium on Flexible Automation, July 1996

3) Michihiko Hayashi, Takashi Tomii, Hiroshi Arisawa: "A Form Acquisition of Subjects in Image Databases", IPSJ SIG Notes Vol.97, No.7, 97-DBS-111-13, 1997

4) H. Arisawa, T. Tomii, H.Yui, H. Ishikawa: "Data Model and Architecture of Multimedia Database for Engineering Applications," IEICE Trans. Inf. & Syst. ,Vol.E78-D No.11, November,1995

5) Kageyuu Noro: "Illustrated Ergonomics", JIS, 1990

6) Kohtaro Fujita: "Jintai-Kaibougaku", Nankoudo, 1993

7) R. Nakamura, H. Saitoh: "Foundmental Kinematics", Ishiyakushuppan Cop,1995

8) M. Hirose, "Image-Based Virtural World Generation," *IEEE Multimedia*, Jan-Mar 1997, pp.27-33(1997).

9) T. Kanade, P. Rander and P. J. Narayanan, "Virtualized Reality: Constructing Virtual Worlds from Real Scenes," *IEEE Multimedia*, Jan-Mar 1997, pp.34-47(1997).

10) S. Kuroki, K. Ishizuka and A. Makinouchi: "Expressions of Spatio-Temporal Queries in the Spatio-Temporal Database System Hawks", IPSJ SIG Notes, Vol.97, No64, 97-DBS-113-58, pp.347-352, 1997

11) T. Teraoka, M. Maruyama, Y. Nakamura, S. Nishida: "The Multidimensional Persistent Tree: A Spatio-Temporal Data Management Structure Sutable for Spatial Search", IEICE, Vol.J78-D-II, No.9, pp.1346-1355, 1995

12) J. A. Orenstein and F. A. Manola, "PROBE Spatial Data Modeling and Query Processing in an Image" Database Application, *IEEE Trans. on Soft. Eng.*, Vol.14, No.5, pp.611-629 (1988).

13) T. Tomii, K. Salev, S. Imai and H. Arisawa, "Human Modelling and Design of Spatio-Temporal Queries on 3D Video Database," Y. Ioannidis and W. Klas eds., *Proc. of the IFIP 2.6 Working Conference VISUAL DATABASE SYSTEMS-4 (VDB-4)*, Chapman & Hall, London, pp.317-336 (1998.5).

14) M. J. Egenhofer, "Spatial SQL: A Query and Presentation Language," *IEEE Trans. on Data and Eng.*, Vol.6, No.1, pp.86-95 (1994).

15) S.Imai, K.Salev, T.Tomii, H.Arisawa: "Modelling of Working Processes and Working Simulation based on Info-Ergonomicas and Real World Database Concept", Proc. of 1998 Japan-U.S.A Symposium on Flexible Automation, pp.147-154, July, 1998

16) H. Arisawa, T. Tomii, H. Yui and H. Ishikawa, "Data Model and Architecture of Multimedia Database for Engineering Applications," *IEICE Trans. Inf. and Syst.*, Vol.E78-D, No.11, pp.1362-1368(1995).

17) P. P. Chen, "The Entity-Relational Model Toward a Unified View of Data," ACM Trans. on Database Systems, Vol.1, No.1, pp.1-49 (1976).

18) H.Arisawa, T.Tomii, K.Salev : "Design of Multimedia Database and a Query Language for Video Image Data", International Conference on Multimedia Conputing and Systems, IEEE, June 1996

8

Spatio-temporal Browsing for Video Databases

Masatoshi Arikawa

Center for Spatial Information Science
The University of Tokyo

ABSTRACT

Time-series spatial description data concerning a camera's movement are expected to be automatically generated by various sensors in the near future. Such description data as a camera's time-series positions, directions and zoom ratios will provide a rich environment for retrieving and browsing video data spatially. Real- time 3D CG (three dimensional computer graphics) is used for user interfaces to browse videos in a virtual space corresponding to an existent space in the real world. Cameras' movements or video sequences are represented as 3D icons in the virtual space. If we click one of the 3D icons, the corresponding video sequence will be replayed in the virtual space. This paper presents a basic principle of the 3D spatial hypermedia for video data browsing, and shows some demonstrations of our prototype system. The basic procedure for producing virtual spaces dynamically is realized by dynamically retrieving and visualizing spatial data stored in spatial databases using the rule of *LoD* (*Levels of Detail*). In addition to the spatial data, *time* data automatically generated as description data of the video data are useful for video retrieval and structuring. This paper introduces a new concept *time walk-through* for retrieving video data using *time* dimension. The concept enables users to travel time in a virtual space. The time walk-through is based on "time" extension to LoD. Our prototype system "TimeWalk" based on our proposed framework is explained with real examples.

1 Introduction

There is a large amount of video data produced by electronic consumer products such as video cameras and personal computers. Finding parts of our interests from a large amount of video data becomes a big problem. Some researchers have tackled the problem using various technical methods, such as keyword searches, natural language processing[5)] and pattern recognition[2)]. As an example of keyword searches and natural language processing, all news programs on TV in the United States are required to have a channel that provides superimposed dialogue for hearing-disabled people. The superimposed dialogue is useful to find parts of our interests from news programs because video contents correspond to their superimposed dialogue. As an example of pattern recognition, there is another research concerning identifying people using image and voice recognition. Thus, the description of videos is used for structuring a large amount of video data. It allows users to retrieve collections of video frames of their interests by means of retrieving description data instead of video data themselves. Text data corresponding to sound data of video data are usually created by hand. It is expected that the image and voice recognition becomes feasible for automatically generating the text data corresponding to the video data, but the quality of the recognition is not good enough at present.

Spatial sensors such as GPS (Global Positioning System) and gyro will become affordable. When the spatial sensors are used with video cameras, the positions and directions of the cameras can be automatically generated as the *spatial description data* of video data. Such spatial description data are useful for retrieving video data[1)], since the spatial description data are generally cheap and reliable. This paper proposes a new framework of video data retrieval using spatial description data and 3D visualization. In addition to the spatial data, *time* data are automatically generated as description data of video data. The time data is also useful for video retrieval and structuring. This paper introduces a new concept *time walk-through* for retrieving video data using *time* dimension based on the "time" extension of the concept *LoD* (*Levels of Detail*). The new concept *temporal LoD* enables users to travel time in a virtual space.

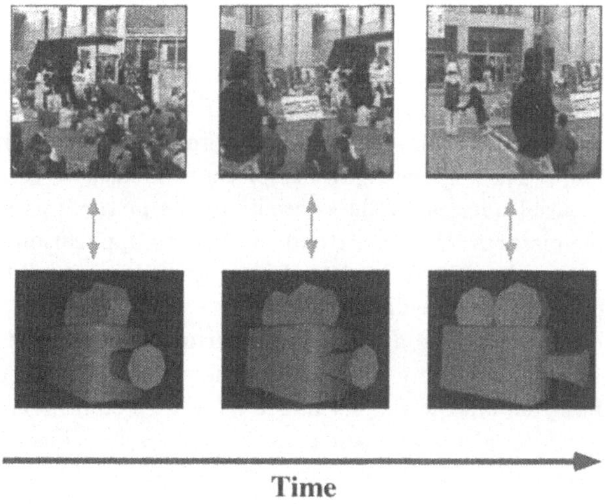

Figure 1: A sequence of time-series video frames and camera movements in the real world

2 An Overview of Spatial Browsing for Video Data

This chapter overviews the use of spatial description data of video data for browsing and retrieving video data of users' interests using spatial queries. A video data is considered a collection of time-series images in this paper. First, we discuss the use of some spatial description data of *photo pictures*, that is, images, then extend it to the use for videos later. In the remainder of this paper, a picture means a photo picture which is one image. One picture corresponds to a camera at a certain moment. The camera at a certain moment is represented by some spatial attributes, such as its position, direction, zoom ratio and so on. 2D (dimensional) map data are also used for making their relations to some spatial attribute data of the camera. If the time-series positions of the camera for some duration are visualized on a 2D map, the visualization can show the distribution of the camera's movement which corresponds to a collection of time-series pictures. The position data can also be used to create clickable icons representing time-series pictures. If we click one of the position icons, the picture corresponding to the clicked icon will be shown on a screen. Also, we use the direction and positi on of the camera to represent cameras or pictures. In this case, the camera can be visualized as an arrow which provides the information on the direction and position. The arrow on the 2D map enables users to understand which direction's scene can be viewed in the picture. If information about the position, direction and zoom ratio of the camera are available, we can tell what region in the real world was taken or viewed in a picture at a certain moment. The region in the 2D map can also be used as a clickable icon for users' interactions. If we click the clickable region icon, the corresponding picture will be displayed on a screen.

The spatial data, such as position and direction of a camera, can be used for spatial queries. The spatial data are also used as clickable icons for 2D map hypermedia as we mentioned before. For example, if we want some pictures of Mt. Fuji, we can realize it by making a spatial query to find regions containing the point representing Mt. Fuji. The regions represent pictures. Thus, we can indirectly retrieve pictures by spatial queries on the contents in the pictures using spatial data corresponding to the pictures. We can extend this idea to applications of video data. Video data can be considered a collection of images or pictures. The unit images comprising a video sequence are called *frames* of the video sequence. Each frame of the video sequence corresponds to a camera at a certain moment and it has its spatial description information (Figure 1). Figure 1 shows the correspondence between a camera's momentary condition and its image as well as some temporal relations of time-series images. We also make some spatial queries for video sequences. For example, we can make a query to select some video sequences which show Mt. Fuji. For cameras taking videos, their time-series positions and directions are expected to be automatically recorded by using spatial sensors in the near future. We can see the camera's movement and replay the corresponding video at the same time. If we visualize all spatial description data for all frames of a video sequence as arrow icons or region icons on a 2D map, the number of arrow or region icons becomes too large to browse and click. We must simplify such many icons for video sequences. For

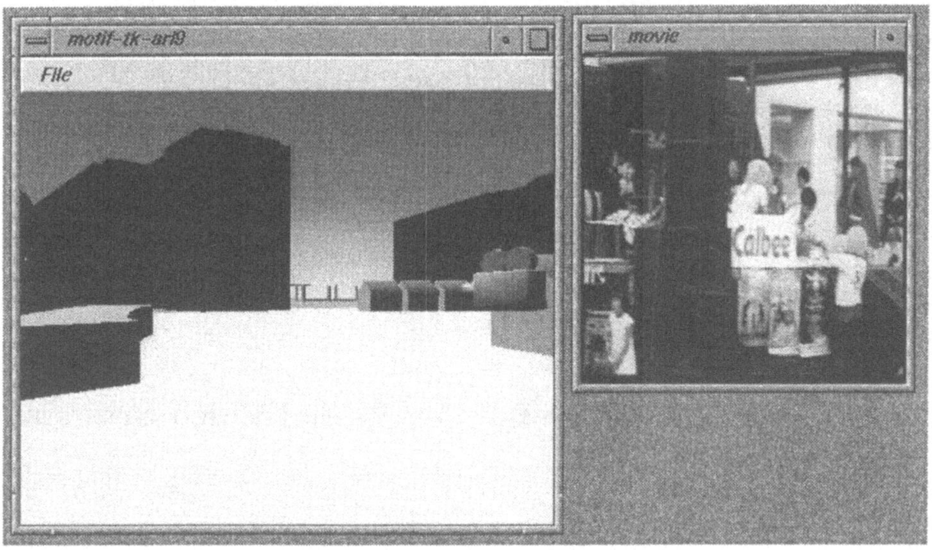

Figure 2: A moving camera in a 3D virtual space and the corresponding video replayed in another window

example, many arrow icons can be represented by a small number of arrow icons. The representative arrow icon represents video sequences or a camera's movement for a certain duration. A video sequence is replayed when the corresponding icon is clicked by a user.

There is a problem of how to divide a video sequence into multiple video sub-sequences which are represented by icons. For example, each arrow icon may represent a segment of every 5-minute video sequence. It is useful to divide a video sequence into more meaningful sub-sequences, but it will be much more difficult than dividing them into pieces of constant duration sub-sequences. A simple method of generating a representative arrow icon for a video sequence is to use the average values of the positions and directions of a set of time-series momentary cameras. The method often fails because it is generally difficult that an average value represents all values for any case. Methods of generating representative arrow icons should be appropriately selected for various cases.

We can extend the idea of spatial browsing of videos on a 2D map to the one for a 3D CG space. All 3D CG spaces discussed in this paper correspond to existent spaces in the real world. Such 3D CG spaces corresponding to parts of the real world are called *3D virtual spaces* in this paper. We walk through a 3D virtual space, and browse or retrieve our intended video sequences which look similar to the current views in the current 3D virtual space. These requests of interactions from users can be interpreted as spatial queries which find arrows or regions closer to the user's intention. The view in the 3D virtual space means the user's intended region and can serve as the selective condition of a spatial query. For instance, while walking through a virtual space, a user can click a 3D arrow icon representing a video sequence so as to replay it in another window on a screen (Figure 2). Furthermore, it is possible to incorporate replaying videos into the 3D virtual space as components of the virtual space (Figure 3, 5). Users can appreciate past real-world videos in the 3D virtual space. This kind of application is called *Augmented Virtuality*[4]. It provides users with more spatial experience. Another application of using spatial queries for 3D virtual spaces is to retrieve and replay videos which show a user's clicked objects in the 3 D virtual space. This application uses a spatial query to select regions, which correspond to video sequences, including or intersecting the user's clicked objects.

3 An Experimental System for Spatial Video Browsing

For our experiment, we used two sensors, a GPS and a gyro to generate time-series spatial data for video sequences which were taken by a digital video camera. We collected some spatial data for videos taking some scenes of campus festival of Hiroshima City University held in October 18th and 19th, 1997. The gyro was precise enough to record the direction of the camera, but the GPS was not precise to measure the position of the camera. We compensated the position data of the camera by plotting every one second positions of it on a 2D campus map by hands later. The zoom ratio can be obtained automatically from a digital camera of the latest model. To simulate the continuous movement of the camera as real-time 3D computer graphics animation, we used a method of linear interpolation for the discrete time-series spatial data such as position, direction and zoom ratio.

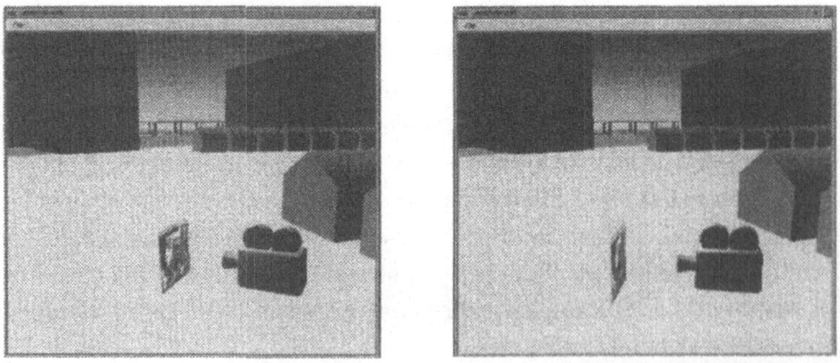

(A) At the time "t" (B) At the time "t+Δt"

Figure 3: Replaying a video on a rectangle plane with a moving camera icon in a 3D virtual space

We have implemented two typical applications for spatial browsing video data.

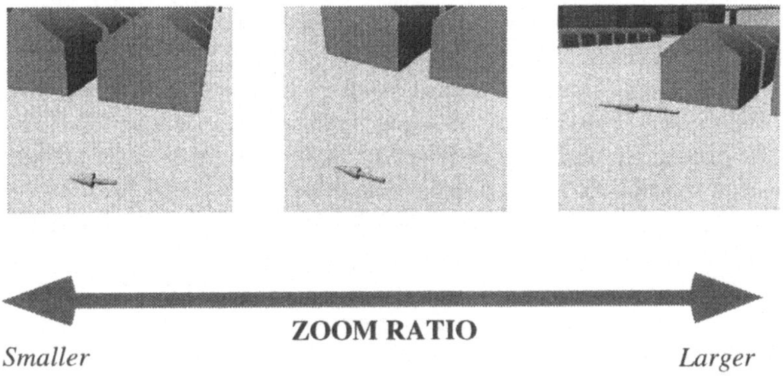

Figure 4: Zoom ratio of a camera is represented as the length of a 3D arrow icon

Figure 5: A replaying video rectangle plane positioned in the center of a window as a result of synchronization of a virtual camera and a real camera

One application is to simulate the movement of the camera as real-time 3D CG animated camera icons in a virtual space. Users could appreciate the movement of 3D CG animated camera icons in the virtual space from arbitrary viewpoints. If we click a 3D CG animated camera icon at a moment, the video corresponding to the current 3D CG animated camera icon is replayed on a CG rectangle plane appearing in front of the 3D CG animated camera icon in a virtual space (Figure 3). The zoom ratio of the camera is used for determining the distance from the CG rectangle plane to the CG camera for replaying a video. Thus, we can appreciate spatially both the movement of the camera and replaying videos in a 3D virtual space.

The other application is to use 3D CG still arrows, each of which represents each segment of video sequences. In the experiment, a video sequence was divided into meaningful segments of video sub-sequences by hands. 3D CG still arrow icons representing the segments of the video sequences were automatically generated by visualizing the average values of the position, direction and zoom ratio of the camera. The length of the arrow represents the average value of the zoom ratio of the camera (Figure 4). We can walk through the 3D CG still arrow icons which address video sequences and indicate the average values of the position, direction and zoom ratio of the movement of the cameras. If we click one of the arrow icons, we can appreciate replaying videos in a virtual space (Figure 5). In this application, the camera to view the virtual space is fixed in the same position, direction and zoom ratio of the real-world camera. We are guaranteed to see the video in the middle of the scene in the virtual space. Compared with the previous application, we appreciate a video from a right angle, but the position and angle of our view cannot be changed in this application. Furthermore, we could have a wider view compared to only a video being played in another window on a screen. The video can be augmented as a wider view and be imposed on a virtual space so that users can experience the video more spatially.

4 Frame Grouping

Since spatial data, that is, the position and direction describing the movement of a camera are available, video data can be clustered depending on the movement of the camera. We can tell different scenes using the discontinuity of the movement of the camera. Even if the position of the camera is continuous, the movement of the camera is clustered by means of its acceleration. For example, a camera translates, then it stops and stays at a fixed point, then starts rotating. In this example, the movement of the camera is divided into three phases: translation, stay, and rotation. In order to cut some scenes, we introduce an algorithm which compares successive video frames for grouping them. Atomic component of the video data is a video frame. Scene cutting is realized by grouping video frames into some scenes. Figure 6 shows the results of frame grouping using our proposed algorithm. The upper bar is the result using the recognition of humans. The lower bar is the result by our algorithm. The two bars show that our algorithm works well.

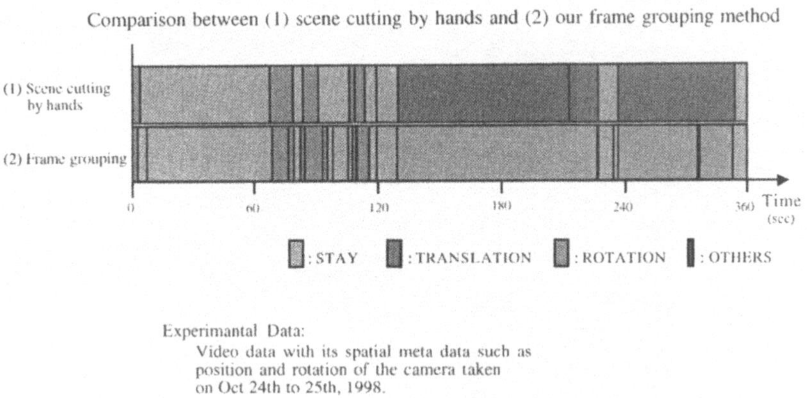

Figure 6: An example for evaluating our frame grouping method

Figure 7: Visualizing "all frames" as "one kind" of 3D "arrow" icons

Figure 8: Visualizing "groups of frames" as "three kinds" of 3D "arrow" icons

5 Visualizing Time-Interval Objects as 3D Arrow Icons

We have researched on browsing video data in virtual spaces. As described in the previous section, each video frame has its spatial description data such as position and direction of the camera. are visualized using only its position data. Figure 7 presents an example of visualization of all video frames as 3D arrow icons using its direction as well as its position. This case is not practical because the number of polygons representing the movement of the camera is too large for users to walk through a virtual space on current computers. The number of video frames per second should be 60 for smooth walk-through. It is important to simplify the representation of the movement of the camera in order to deal with a large amount of video data. Our proposed frame grouping is useful for simplifying the representation of the movement of the camera. One 3D arrow icon can represent a group of multiple video data. For example, a representative 3D arrow icon is created using averages of both positions and directions for all video frames belonging to a frame group. frame groups by one kind of 3D arrow icons. Figure 8 shows an example of using three kinds of 3D arrow icons which correspond to three kinds of frame grouping: translation, stay and rotation. It is convenient for users to find out video data of their interests in a virtual space with smooth walk-through.

6 Spatio-Temporal LoD

Our proposed frame grouping can decrease the number of representative 3D arrow icons as visualization of the movement of the camera. For longer duration, too many 3D arrow icons will appear in a virtual space. It is necessary to provide users with some tools or equipment for setting intended time and duration easily as the condition of selecting spatial objects from spatial databases. The number of polygons of all 3D arrow icons must be below a constant number, for example, 10,000, in order to walk through a virtual space smoothly.

LoD (*Levels of Detail*) is the main rule for realizing walk-through based on spatial databases. The rule is used for selecting spatial objects from spatial databases by spatial queries, and for constructing a virtual space dynamically. We extend the LoD in order to take *time* into account for calculating the distance between a user and objects. The time extension for LoD leads us to *time walk-through*.

6.1 LoD

LoD is determined depending on the positions of a camera and these of some objects in a virtual space. For example, a frame group is visualized as multiple arrows, one arrow, only point, and none, depending on the distance between the camera of the frame group and the viewpoint of a user (Figure 9). When a user sees the camera at a near position, the precise representation of its movement is adopted. If a user is far from the camera, the representation of its movement becomes rough. This way of controlling the quality of spatial data is natural and effective for practical walk-through.

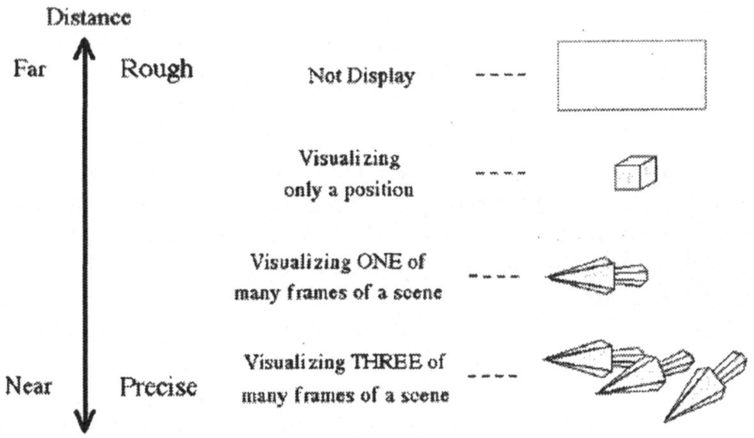

Figure 9: LoD for space and time scale distance

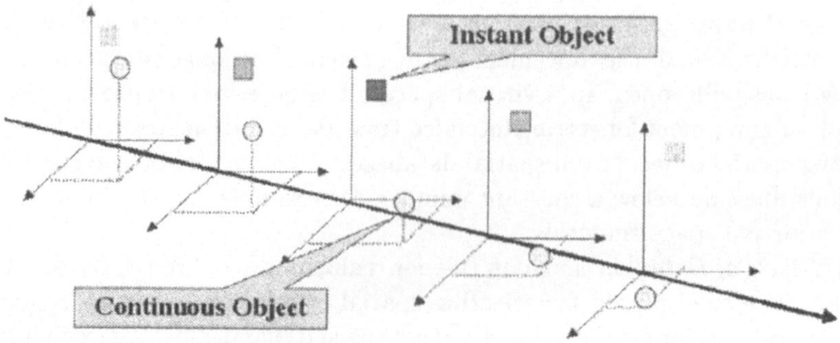

Figure 10: Images of time walk-through with continuous and instant objects. A user is moving on a time axis or is getting closer to a particular time. An instant object existing at only a moment is going to appear or disappear.

6.2 Concepts of Time Walk-Through

In the case of ordinary virtual spaces whose time is usually fixed or is a point on the time axis, the quality of spatial objects is controlled depending on the space scale distance without considering the time dimension. When we think about walk-through in a virtual space and *time*, the distance should be defined by the time scale distance as well as the space scale distance. Near objects mean objects which are near in the time dimension (t) as well as in the spatial dimensions (x, y, z). If we realize a time extension to LoD, we can walk through on the time axis. It may be difficult to imagine the time walk-through. We show some examples to give a clear image of the time walk-through in Figure 10. In the real world, we see objects on the latest point on the time dimension. When we watch a past video, we only see a point on the time dimension. Real world objects are usually continuous on the time dimension as long as they exist. On the other hand, there are instant objects that exist on points in the time dimension. For example, a photo is a point object on the time dimension, because the time attribute of the photo can be defined by a point on the time dimension. To realize an ordinary virtual space using a region query applied to spatial databases, the condition of the time attribute may be a point on the time dimension. It works for retrieving continuous objects but not for retrieving instant objects by temporal point queries. Intersections between a point of a query and a point of a spatial object on the time dimension are generally useless. In order to deal with instant objects, we need to convert either a point as a query's condition or a point as a spatial object's time attribute into an interval of time. When a user is coming closer to an instant object on the time dimension, the temporal LoD of the instant object is increasing, and the user can obtain more precise visualization of the instant object.

6.3 Correspondence between Space and Time

From a technical point of view, it is easy to extend the concept of LoD for 3D virtual spaces to 4D virtual spaces including the time axis, but it may be difficult to clarify the meanings of the time extension to the LoD. We will try to clarify the concept of the time extension to LoD using correspondence between space and time. The viewpoint in a virtual space corresponds to the *present time* on the time axis. The walk-through in a virtual space is realized by the movement of the viewpoint, while the time walk- through is realized by the movement of the present time. Considering information retrieval, a retrieving area of the view for a walk-through in an ordinary virtual space corresponds to a time interval whose center is the present time. In the case of the LoD for 3D spaces, the LoD of each object in the view is determined by the distance from the camera to the object. On the other hand, in the case of the LoD for the time dimension, the LoD of each object is determined by the time scale distance. If a user comes close to an exact time in a virtual space, near objects appear as detailed representations, and far objects disappear. By changing the present time, users can walk through a virtual space over time.

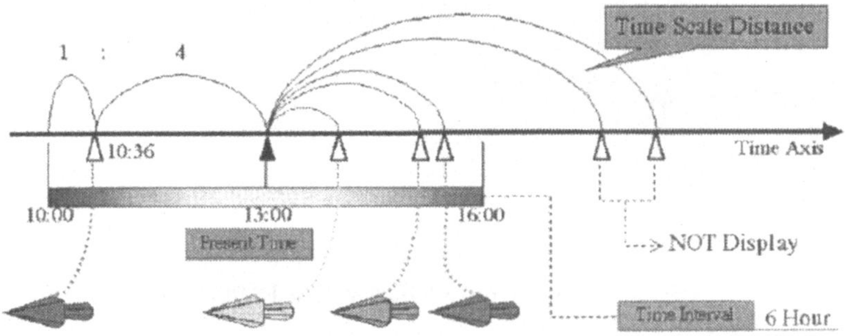

Figure 11: Arrow's color is decided by when an object exist on the time interval

6.4 Color Decision by Time Scale Distance

In addition to the LoD for the time dimension, it is useful to change the color of a spatial object according to the ratio of the exact time of the object's existence to the time interval. This section shows examples of color decision by time scale distance using Figure 11. An object exists at 10:36, the present time is 13:00, and the present time interval is 6 hours. In other words, a user can see all objects which exist from 10:00 to 16:00 in the virtual space. The object at 10:36 is also visualized in the virtual space. The color of the object is blue because the ratio to the time interval is 1:4 to blue:yellow. Other instant objects have their own colors according to the time of each object. In the example, the objects near to the present time have yellow as their colors. The colors of objects which are behind the present time are blue, and the colors of those which is ahead of the present time are red. Users can find out the rough time of spatial objects from their color. If a user moves to the advancing time, the blue objects would change into yellow and red successively. Also, if a user changes the time interval into a longer one, the number of objects existing in the time interval increases and the colors of the objects change according to the ratio of the time of each spatial object to the present time interval.

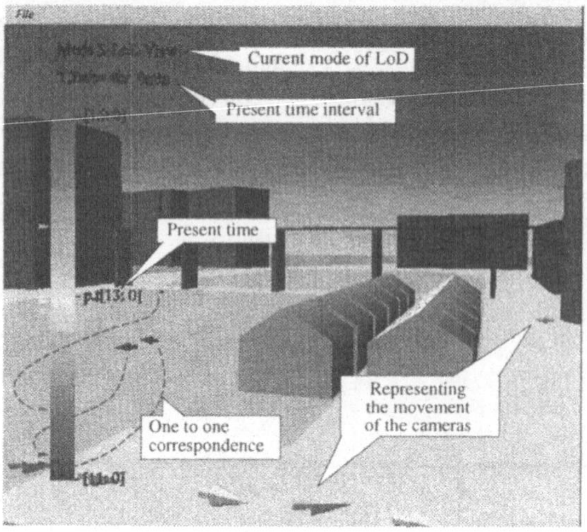

Figure 12: Time bar with a virtual space
(Present time, Present interval)=(13:00,4:00)

6.5 Prototype System "TimeWalk"

We implemented a prototype system called *TimeWalk* to realize the time extension to LoD. We used an O2 workstation of SGI (Silicon Graphics Inc.) and SGI IRIS Performer 2.0 graphics library for prototyping. Our experimental spatio-temporal data are the movements of video cameras. The multiple representations of the spatio-temporal objects are a point, an arrow and multiple arrows for LoD. There are three modes of LoD for our prototype system: S-LoD (LoD by Space scale distance), T-LoD (LoD by Time scale distance) and ST-LoD (LoD by Spatio-Temporal scale distance).

We explain the user interface of the prototype system using Figure 12. The 3D arrow icons represent the movements of the camera. There is the bar called *time bar* in the left side of the display screen. The time bar represents the present time interval when all visible 3D arrow icons exist. In this example, the time bar represents

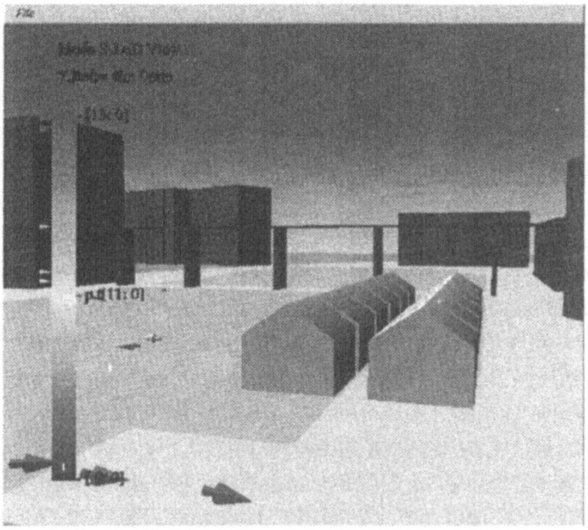

Figure 13: Changing the present time
(Present time, Present interval)=(11:00,4:00)

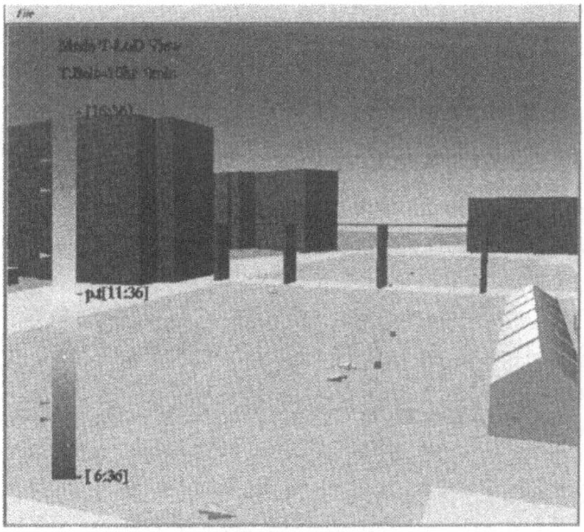

Figure 14: Changing the LoD mode into "temporal" LoD

4 hours, that is, from 11:00 to 15:00. The present time 13:00 of the virtual space is represented in the center of the time bar. There are some small triangles on the left side of the time bar. The triangles represent the time of spatio-temporal objects, that is, cameras as points in the time axis. Thus, the virtual space is a visualization of spatial attributes of spatio-temporal objects, while the time bar is a visualization of temporal attributes of spatio-temporal objects. A user can change the present time and the present time interval of the time bar. Changing the present time realizes time walk-through, and corresponds to changing the viewpoint or walking through a virtual space. Changing the present time interval may correspond to changing the field of view in the virtual space. There is another interpretation that extending the present time interval means going backward while keeping the central point fixed in the virtual space for extending the view. Thus, changing the present time interval is related to changing the scale of time. We can also select the mode for LoD from the selections: Spatial LoD (S-LoD), Temporal LoD (T-LoD) and Spatio-Temporal LoD (ST-LoD). The three kinds of LoD are defined by different definitions of the distance: 3D distance, time distance and 4D distance. Figure 13 shows changing the present time, that is, going back to a 2 hours past virtual space. Yellow arrows are changed into red arrows because they exist in the future from the present time. interval. Extending the present time interval increases the number of objects in the view, all objects become yellow. Shortening the present time interval decreases the number of objects. Figure 14 shows changing the LoD mode into Temporal LoD.

7 Conclusions

Spatial sensors are expected to be affordable and become precise enough to generate spatial description data for objects in the real world in the near future. The spatial description data of a video camera's condition measured by the spatial sensors are also useful for browsing and retrieving video data spatially. The spatial data of the video camera provide richer visual index environment. We can retrieve both pictures and videos in 3D virtual spaces as well as on 2D maps. Also, real-time 3D computer graphics hardware enables us to walk through a 3D virtual space and to appreciate videos replayed in the 3D virtual space. This kind of environment can provide users with spatially browsing of videos in a virtual space. We can experience virtual spaces in more realistic ways because real world videos can be replayed in a virtual space. While we walk though 3D virtual spaces which have their corresponding existent real world spaces, we can be aware of the existence of videos in t he virtual spaces because there are 3D icons for video sequences, and can appreciate replaying videos more spatially and naturally in the virtual spaces. This application promises to manage video data for certain application domains, such as sightseeing tours of the real world, simulations of the real world, and virtual disaster management systems.

This paper has presented a concept *Time walk-through* by means of introducing time extension to the ordinary LoD. This framework is significant for browsing spatio-temporal data such as video data in a virtual space. We implemented a prototype system and proved that it worked well. The framework of the time walk-through is feasible for a wide variety of applications, including temporal transition of real world objects such as buildings, spatio-temporal reference to historical records, and biological evolution and change of habitat.

Acknowledgments

I would like to thank Mr. Tetsu Kamiyama, Mr. Takashi Maesako and Mr. Takashi Sueda for their cooperation in this research as parts of their graduation theses of Faculty of Information Sciences, Hiroshima City University. This research was supported in part by the Grant-in-Aid for Scientific Research on Priority Areas "Advanced Databases" of the Ministry of Education, Science, Sports and Culture of Japan, and "Research for the Future" Program of Japan Society for the Promotion of Science under the Project "Advanced Multimedia Contents Processing" (Project No. JSPS-RFTF97P00501).

Bibliography

1) M. Arikawa and T. Kamiyama, "Spatial Browsing for Video Databases," Proc. the 1st Int'l Conf. on Advanced Multimedia Contents Processing (AMCP'98), 1998, Shojiro Nishio and Fumio Kishino Eds., Springer, Lecture Notes in Computer Science, 1554, pp. 313–327.

Video Databases", Environments (DANTE '99) ,

2) Y.Ariki, A.Shibutani and Y.Sugiyama "Classification and Retrieval of TV Sports News by DCT Features," IPSJ97, 1997, pp. 269–272.

3) ART+COM, http://www.artcom.de/

4) Milgram, P., Kishino, F.: A Taxonomy of Mixed Reality Visual Displays, IEICE Transactions of Information Systems, Vol. E77-D, No. 12 (1994) pp. 1321–1329

5) K. Uehara, M. Oe, K. Maehara, "Knowledge Representation, Concept Acquisition and Retrieval of Video Data," Proceedings of the International Symposium on Cooperative Database Systems for Advanced Application, Kyoto, 1996, pp. 218–225.

9

GIS Infrastructure in Japan — Developments and Algorithmic Researches

Hiroshi Imai
University of Tokyo

Keiko Imai
Chuo University

Kazuo Inaba
Geographical Survey Institute

Koichi Kubota
Chuo University

ABSTRACT

Some activities of geographical data maintenance and standardization work in GIS infrastructures in Japan are first reviewed. Then, from the viewpoint of geometric and combinatorial algorithmics, applied aspects of GIS are described. Some spatial data mining technique is mentioned first, and then the problem of inferring topological information from digital map and the label placement problem are touched upon. As one of topics in ITS (Intelligent Transport System), finding useful detours in car navigation is discussed. These would illustrate some advanced aspects of GIS infrastructure and its high-level use in Japan.

1 Introduction

Geographical Information System (GIS) has two aspects: one is to create and maintain geographical data in digital form, and the other is to provide efficient ways of utilizing geographical data for various purposes based on the information technology. Concerning the former aspect, in these two decades, big efforts have been made to construct digital geographical data as GIS infrastructure. Traditionally, base geographical information is given as maps, available only on papers, which were hard to handle directly on computers. With the evolution of digital geographical data, we are now at a high stage of using these data interoperably via networks throughout the world in a variety of fields. Here, standardization of GIS data plays a crucial role, and in fact several important standardization activities have been performed. In the first half of this paper, we survey GIS developments in Japan for maintaining geographical data and for contributions in standardization work. Some of activities of a representative governmental institute performing research and servicing on geographical data are described, together with some of typical digital geographical data maintained so far in Japan.

One of the most frequently used GIS systems in Japan would be the car navigation system. For example, dynamical traffic information newly becomes available such as ATIS (Advanced Traffic Information Service) and VICS (Vehicle Information & Communication System) besides compact car navigation systems attached to vehicles, based on the GIS infrastructure. In the latter half of this paper, we describe some of research results on GIS from the standpoint of spatial data mining, computational geometry, and network algorithms. These include demonstration of a k-means algorithm for spatial data mining, the problem of inferring topological information from digital map, and the label placement problem for node and edge features in maps. Also, as an interesting issue in ITS (Intelligent Transport System), the problem of finding useful detours in car navigation is discussed.

By covering the two aspects of GIS in the above-mentioned way, this paper tries to give outline of some of GIS infrastructure and GIS research in Japan.

2 GIS Developments in Japan

In this section, recent GIS-related activities in Japan are described. The section covers three subjects. The first subject is digital map preparation, which was initiated about 25 years ago. At that time, there was no such concept as spatial database, and efficient map digitization was a main research problem. The second subject topic is recent GIS-related activities by the Japanese government. Information on current activities and the policy of Japanese government are presented. The third topic is about the Spatial Data Framework, which was developed under the new concept of spatial database.

2.1 Development and Publication of Map Data

2.1.1 Outline

The development process of GIS infrastructure in Japan can be explained in four phases. Phase I began in the middle of the 1970s when the government started preparation of digital geographic data for only limited users such as central and local governmental organizations and researchers at universities. Phase II arrived when the Geographical Survey Institute (GSI), Ministry of Construction (MOC) began to publish digital cartographic data sets in 1993. Phase III started in 1995 when the government reached a consensus that the active encouragement of GIS development was necessary. At present, Phase IV is progressing when the preparation of spatial database in accordance with a standard is important. Below, phases I and II are explained. Phase III is explained in section 2.2.

2.1.2 Development of Digital National Land Information

The Japanese government has been developing digital geographic information since the mid-1970s. As its initial activity, GSI began to develop the "Digital National Land Information" in 1974 in cooperation with the National Land Agency (NLA), and was nearly completed in 1980. Its accuracy corresponds to approximately 1:200000 paper maps. It consists of DEM, land-use data, boundaries of local governments, major roads, railways, rivers, coastal lines, public facilities, etc. The purpose of this project was to supply basic digital geographic data necessary for national land development planning and regional planning by the central governmental agencies and local governments.

It has also prepared the "Detailed Digital Land Use Data" to support the policy making of building land administration in collaboration with the Economic Affairs Bureau of MOC since 1981. It is a data set of grid cells for land use (10m square on the ground) for three major metropolitan areas (Tokyo area, Osaka area, and Nagoya area), and each area is surveyed repeatedly every 5 years.

These data sets have been highly reputed for they have enabled quantitative analysis of national land. However, they have been specially prepared for administrative purposes, therefore they have not been disclosed to the public but used only by administrators within the central and local governments and researchers at universities.

2.1.3 Publication of Digital Geographic Information

In June 1993, GSI launched into the publication of digital cartographic data sets called the "Digital Map Series". It was extremely epoch-making. Since then, the variation and number of published digital cartographic data and software that apply those data have increased, and as a result, people have gradually come to recognize the benefits of geographic information.

Nine kinds of "Digital Map Series" are available at present. They are "Digital Map 10000 (total)," "Digital Map 25000 (shore lines and administrative boundaries)," "50m mesh (elevation)," "250m mesh (elevation)," "1km mesh (elevation)," "1km mesh (average elevation)," "Digital Map 25000 (Map Image)," "Digital Map 2500 (Spatial Data Framework)," and "Digital Map 200000 (shore lines and administrative boundaries)". They are text files and distributed via CD-ROM with simple software for quick browsing of the image of inside data.

2.2 Recent GIS-related Activities by the Japanese Government

2.2.1 Liaison Committee of Ministries and Agencies Concerned with GIS

A Liaison Committee of Ministries and Agencies Concerned with GIS was established in September 1995 to promote the efficient development and effective utilization of GIS within the Government with the close cooperation among the Ministries and Agencies. The Cabinet Councilor's Office, Cabinet Secretariat, was designated as the secretariat of the Liaison Committee, and assisted by the Geographical Survey Institute (GSI) and the National Land Agency (NLA). The Committee has two task force groups, i.e. Spatial Data Framework Task Force Group and Basic Spatial Data Task Force Group, each of which has a few working groups to discuss more specific topics in detail.

2.2.2 Long-term Plan for the Development of NSDI in Japan

The Liaison Committee developed a Long-term Plan in 1996 for the development of NSDI (National Spatial Data Infrastructure) in Japan. The Plan specifies actions to be taken by the Government during a two-phase period starting in 1996 up to the beginning of the 21st century. The first phase focuses on the definition of NSDI in Japan as well as standardization of metadata and clarifying the roles of the central government, local governments, and the private sector, rather than actual spatial data development. The implementation of NSDI including spatial data development for NSDI is expected to take place in the second phase. Approximately three years have been designated for each phase, i.e., first phase (1996-99) and second phase (1999-2001).

2.2.3 Pilot Study by Local Governments for the Implementation of the Long-term Plan

The definition of NSDI, one of the main subjects of the first phase of the Long-term Plan, requires intensive research on the availability, utilization, restriction and distribution of maps and spatial data in local governments, because they develop and maintain most of spatial data sets in Japan. GSI and NLA are conducting a collaborative pilot study in fiscal year (FY) 1997 with four local governments to do such research. The main topics of the pilot study include: which spatial data item should be included in the Spatial Data Framework of the Japanese SDI; who should develop and maintain such data items; and which information would be most suitable for indirect georeferencing. The result of this pilot study was summarized in Interim Report of the Long-term Plan at the end of FY 1997. Additional Ministries, i.e., the Ministry of International Trade and Industry (MITI), the Ministry of Posts and Telecommunications (MPT), and the Ministry of Home Affairs (MHA) joined the pilot study in FY 1998 starting in April 1998. The research topics of these Ministries in the pilot study are as follows: MITI will develop new information systems with GIS and foster related industries; MPT will focus on the development of a spatial data search engine through computer networks, spatial data encryption methods to protect private information, and spatial data compression for efficient data distribution; and MHA will investigate technological and institutional issues of local governments related to NSDI development. The results of these pilot studies were incorporated into the final report of the first phase of the Long-term Plan, which will direct the implementation of NSDI during the second phase.

2.2.4 Interim Report on the Implementation of the Long-term Plan

The Committee compiled an interim report on the activities during the first two years of the first phase of the Long-term Plan. The report reviews the subjects specified in the Long-term Plan and the actions actually taken by the Committee. It also clarified the issues to be focused on by the Committee during the last year of the first phase. The report was published and distributed to the public at the end of March 1998.

2.2.5 Final Report of the First Phase

The Committee adopted the Final Report of the First Phase of the Long-term Plan on March 30, 1999. The Final Report entitled "Standards and Development Plan of National Spatial Data Infrastructure"[1]includes two standards of Japanese NSDI (i.e., a technical standard that is based on ISO/TC211 standard drafts, and a list of data items adopted as the framework data) and a development plan for the second phase of the Long-term Plan.

The technical standard included in the Final Report was developed through collaborative research between the Geographical Survey Institute and 53 private companies during the three-year research period starting 1996. Together with these activities, GSI has prepared a new type of digital cartographic data sets called a Spatial Data Framework for city planning area for all of Japan since 1995. The characteristics of these data sets are that they:

1) are structured by several very simple items;
2) distinguish each block as a polygon (suitable for address matching, only for some areas and not for all areas);
3) contain road network structure; and
4) can be used on a personal computer and easily transferred.

The data sources for these files are:

1) data converted from digital maps already held by GSI;
2) newly digitized data from the 1:2500 base map for city planning which local governments keep; or

[1] `http://www.gsi-mc.go.jp/REPORT/GIS-ISO/LCGIS/honbun.pdf` (in Japanese)

3) newly digitized data from the 1:500 map for road management held by some local offices of the MOC.

The data sets have been published from April 1997 for the use of unspecified individuals at an appropriate price, just as the Digital Map Series. They are also distributed free of charge to every local government that provided data sources.

2.3 Other Activities of the Geographical Survey Institute

2.3.1 Research on GIS Standardization

Based on the need to develop a GIS standard for Japan, which is in accordance with that of ISO/TC211, GSI started research on Japanese GIS standard in 1996. This research was also intended to provide a technical backbone for the Japanese SDI standard discussed by the Liaison Committee of the Government. Fifty-three private companies joined this three-year research project funded by the Ministry of Construction as one of the projects of the collaborative research program with the private sector. Two kinds of standards were developed through this research: spatial data exchange standard and spatial data development standard. Six working groups were established for the exchange standard to discuss 8 work items including data structure, data quality, georeferencing, metadata, and cataloguing. Spatial data development standard includes a guideline to develop specifications for spatial data development contracts. The final draft of the standard was developed at the end of FY 1998 and adopted as part of the NSDI Standards by the Government Liaison Committee.

2.3.2 Research on Geographic Information Directory Database (GIDD)

The Long-term Plan developed by the Government pointed out the necessity for National Spatial Data Infrastructure and specifies the need to establish a clearinghouse system for spatial data.

GSI has been developing a Geographic Information Directory Database (GIDD) as a five-year research project since April 1994. This database is designed to provide directory information (i.e., metadata) of spatial data through computer networks, and to become a clearinghouse node by developing a search environment of distributed databases.

The metadata standard, which is currently used in the GIDD, is developed as one of the work items of the Spatial Data Exchange Standard of "Research on GIS Standardization" described above. This standard can be considered as the Japanese

A prototype of GIDD with limited search capability has been in the process of practical testing. This prototype provides 229 records, experimental metadata of Digital Map 10000 (Total) in Japan. GIDD has the ability to make a query using 12 attributes such as title, keyword, longitude, latitude, and producer as well as a combination of these attributes using logical operators. Based on the result of these tests, enhancing its search capability and support distributed environment is planned. A prototype is made public at the GSI homepage.

3 Applications of Computational Geometry to GIS

3.1 Spatial Data Mining in Geographical Databases

Data Mining, or Knowledge Discovery in Databases (KDD) is to find interesting, previously unknown and useful information from large databases. There have been many studies of data mining in relational as well as transaction databases as the first targets of this field. Now, data mining has been extended to other types of databases such as data mining spatial databases, or, spatial data mining [4], and our related results in this setting [6]. Then, we discuss issues to investigate for data mining in geographical databases, especially topological geographical data.

Data mining in spatial databases, or spatial data mining has been proposed; see [4]. Spatial data mining refers to the extraction of implicit knowledge, spatial relations, or other patterns not explicitly stores in spatial databases.

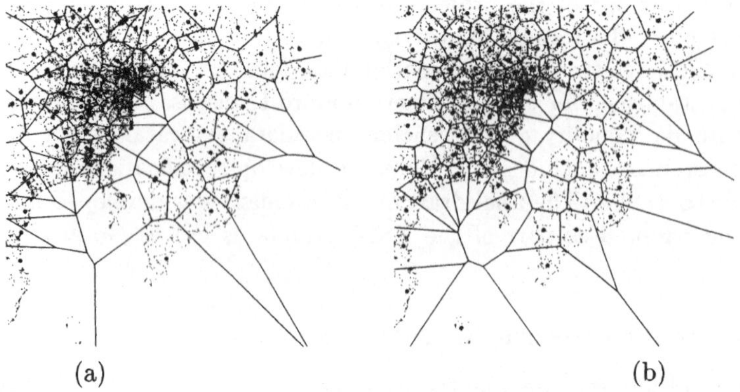

Figure 1: Application of the k-means algorithm for 20726 points at Kanto district in Japan with $k = 100$: (a) initial random solution, (b) solution obtained by the k-mean

In the existing spatial data mining, basically part of geographic information having strong connection with remote sensing, image databases exploration seems to have been investigated, and a clustering approach is adopted to derive knowledge. Main algorithmic tools used in this approach are k-mean, k-medoid and their extensions. The basic algorithms for them are well-known and have been used in many areas. Especially, in connection with geographical databases, the so-called geographical optimization approach provides a general algorithmic framework in terms of mathematical programming and computational geometry [10]. To give an idea about these, we here show an example in Fig.1, taken from [6], of applying the k-mean algorithm to about 20,000 points corresponding to big crossings of road network in Kanto district in Japan. This clustering itself is basically intended for experimental use, and not for some specific data mining, and yet this example would illustrate how

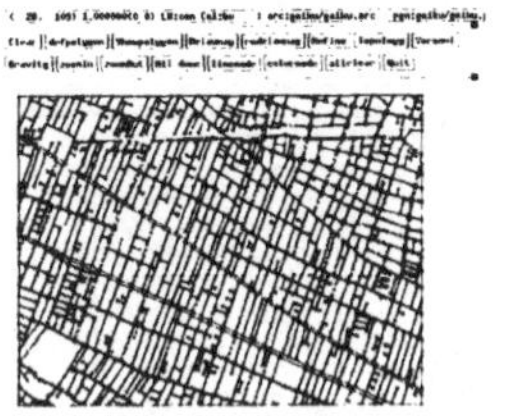

Figure 2: Town map data near JR Nishiogikubo station)

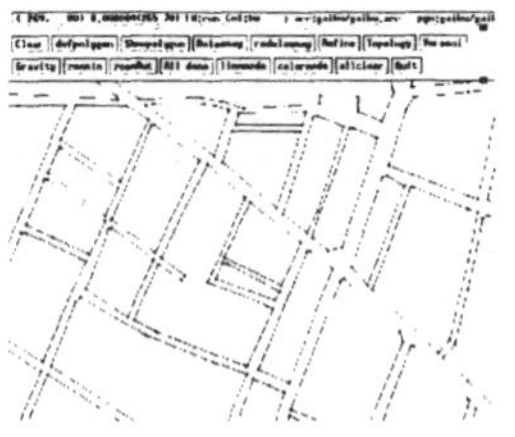

Figure 3: Enlarged map of Town map data in Fig.2 near the south exit of the station)

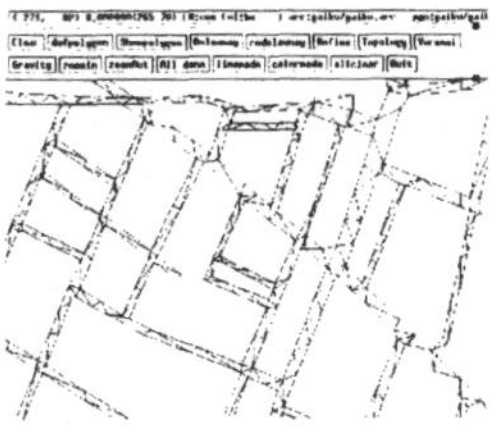

Figure 4: Inferred road network

large the amount of geographical data even in this restricted area and its geometric nature. The *k*-means algorithm work in higher dimensional space, and, general theoretical background is investigated in [7].

3.2 Inferring Topological Information of roads from Map Data

This section describes an application of computational geometry to infer topological information of roads from town map data [13]. Here, an example of town map data is shown in Fig.2. This kind of map data is available as "Digital Map 2500 (Spatial Data Framework)"[2]as described in section 2. In the town map, each town block is represented by (a set of) polygon(s) (Fig.2).

By using these data, enlarging/reducing the size of maps can be performed easily. However, even with these data structures, some sophisticated topological information is not available directly. In order to list all roads incident to a town block or list town blocks incident to a road, topological incidence relations among polygons should be inferred.

Here, by using the above-mentioned Digital Map 2500 as raw data, it is demonstrated that road information can be obtained from only town block boundaries by using computational-geometric algorithms.

Road areas are obtained by erasing town blocks from the map plane. However, with this operation, no topological information about roads is obtained. See a town map near JR Nishiogikubo station south exit of Digital Map 2500 Fig.3. From this data, if the road regions can be triangulated as in Fig.4, we can construct topological network of roads; each triangle represent adjacency.

To derive such a triangulation, Delaunay triangulation and Voronoi diagrams can be used. We here used a program by Prof. Sugihara of University of Tokyo [15] which can handle large-scale geographical data efficiently and correctly.

[2] `http://www.gsi-mc.go.jp/MAP/CD-ROM/cdrom.htm` (in Japanese)

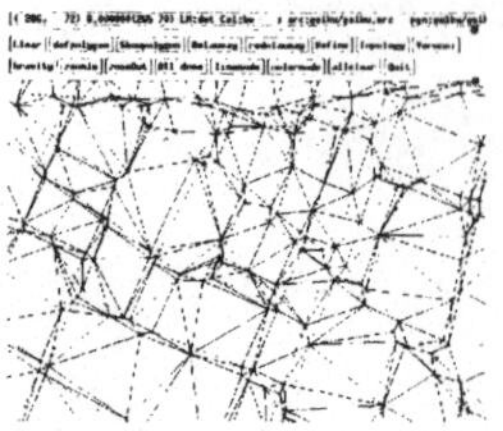

Figure 5: Delaunay triangulation for points of the boundaries

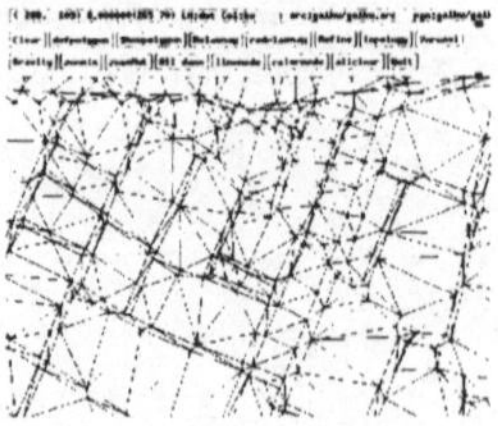

Figure 6: Delaunay triangulation for points densely placed on boundaries

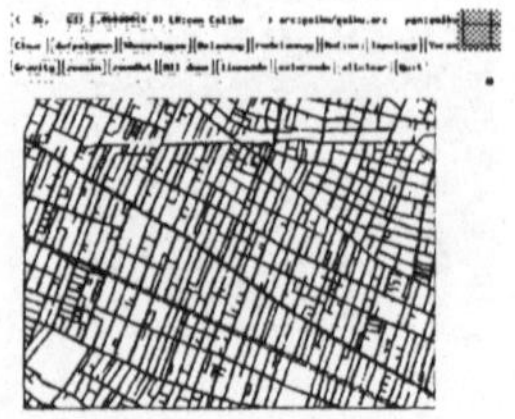

Figure 7: Inferred road network

3.2.1 Topological Inference

To derive a good triangulation as mentioned above, we may use Constrained Delaunay Triangulation (CDT) [2]. However, we here adopt another approach of using Conforming Delaunay Triangulation [3] in order to make full use of the robust and efficient Delaunay triangulation algorithm for point sets.

When we simply compute the Delaunay triangulation for points of polygons in Fig.3, the result looks like Fig.5. In this figure, dotted lines represent a boundary edge of town block which does not become an edge of the Delaunay triangulation. In order to remove such edges, we simply add internal points of such polygon edges so that the resulting Delaunay triangulation for points become conformal to the town block boundaries.

Fig.2 is a town map data consisting of 4168 nodes and 568 town units. Fig.3 gives an enlarged map of some part. For nodes in the map, Fig.6 is a Delaunay triangulation after adding the middle point of each edge which does not become an edge in the Delaunay triangulation for the given nodes (now the number of points increases to 5060).

Starting with the original data in Fig.2 and adding middle points appropriately on the boundaries, we can construct a triangulation of at most twice number of points from which the topological information can be derived. See results in Fig.7 (Fig.4 is its enlarged map).

3.2.2 Inferring Medial Line

From the viewpoint of computational geometry, the above method may be viewed as approximating the Voronoi diagram for polygon edges by the Voronoi diagram of points on the polygon boundaries [15]. With this interpretation, the medial line of roads can be derived directly.

The problem of finding the medial line of road has strong connection with inferring topological structure of roads. A well-know method is to use the Voronoi diagram for boundaries of town unit; then, lines equidistant from adjacent town units are medial lines [9)].

Figure 8: Voronoi diagram when edges are divided into pieces further

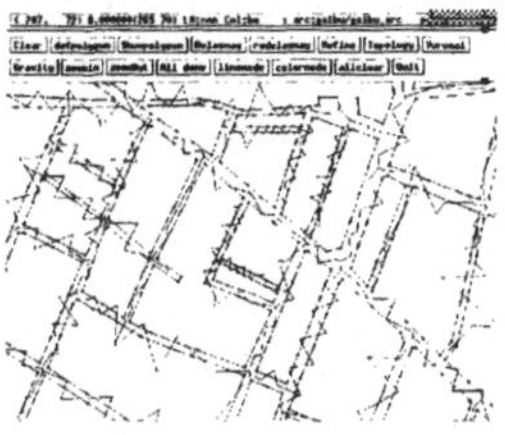

Figure 9: Voronoi diagram

Figure 10: Connecting the centroid and the middle points of triangle edges

The Voronoi diagram for boundary line segments can be approximated by the Voronoi diagram for points densely placed on the boundaries, as depicted in Fig.8, although, when the density is low, edges of the Voronoi diagram may oscillate as in Fig.9. Except crossing points of roads, a method of connecting the middle points of edges of triangles obtained as above may approximate well the Voronoi diagram, which is easier to perform in practice. Practically, we may construct the Voronoi diagram for line segments only near crossing points, and, in other places apply the triangulation method.

3.3 Map Labeling

In maps, labels (or, names) of regions, rivers, and stations, etc., are placed in appropriate places so that the corresponding features in the maps can be understood. Where to place such labels is quite important for readability of maps. The problem of placing such labels are called the map labeling problem, and has been studied intensively [1, 11, 12, 16)]. Here, we describe some recent approaches for edge labeling version of this problem.

The problem of placing a label to a point is call the NLP (Node Label Placement) problem. We can consider may types of candidate positions near the point for placement of the label. Even for the case of fixing the label position uniquely to each point, the problem of placing the maximum number of labels so that no two placed labels do not intersect is NP-hard.

The problem of placing labels for edges is called the ELP (Edge Label Placement) problem. The label for an edge may be placed at any place along the edge, and in this respect there is more freedom than in the NLP problem. One approach to solve this problem is to select a set of points on the edge and place the label of the edge at one of points. Recently, a unified approach of placing labels for edges and nodes simultaneously (called Graphical Feature Label Placement problem; GFLP problem for short) is proposed [12]. The unified approach provides a general algorithm, but, to specific applications, detailed parts of the algorithm should be newly determined. Imai and Kameda [8] propose generalized algorithms for train maps, etc. We here outline these algorithms and show some computational results.

The unified approach for the GFLP problem in [12] consists of the following three steps.

1. Determining a finite set of candidate places for the label to each feature (feature is either a point or an edge)

2. Define the cost of each candidate place.

3. Find a minimum-cost set of candidate places, at most one for each feature, which do not intersect each other.

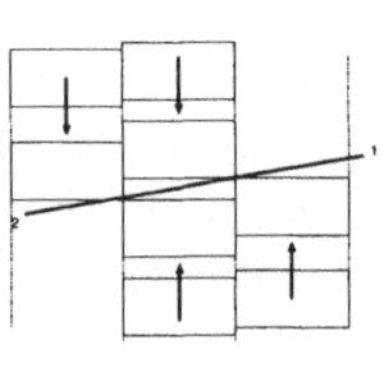

Figure 11: Candidate places of a label for an edge (vertical division case)

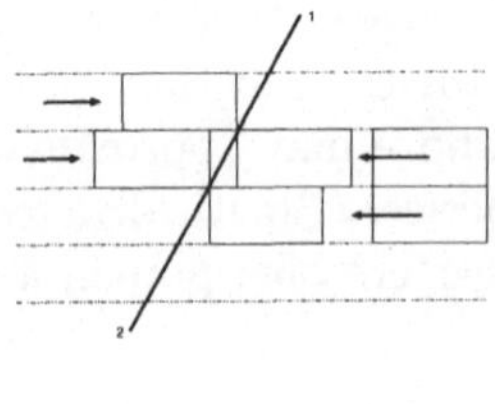

Figure 12: Candidate places of a label for an edge (horizontal division case)

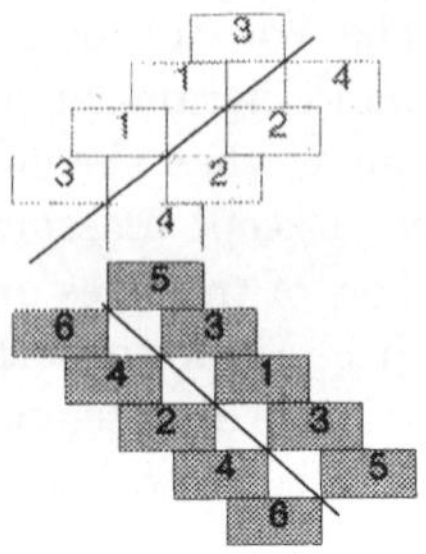

Figure 13: Costs of candidate places of a label for an edge

In order to apply this general framework, we have to determine how to design candidate places for each label, and further how to set the cost of each candidate place. Imai and Kameda's approach [8] works as follows. Suppose that an edge is given as a line segment. For simplicity the size of each label is fixed. Denoting the width and height of an edge by W and H, we compare the slope s of the edge with H/W, and when

$$-\frac{H}{W} \leq s \leq \frac{H}{W},$$

we divide by vertical lines of interval W; otherwise by horizontal lines of interval H, so that a label position incident to the edge is considered as candidate places, as in Fig.11 and Fig.12, respectively. Candidate places intersecting with other edges are deleted. This extends the method in [11].

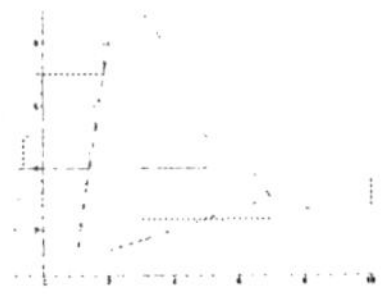

Figure 14: Computed candidate places by the cost-first method

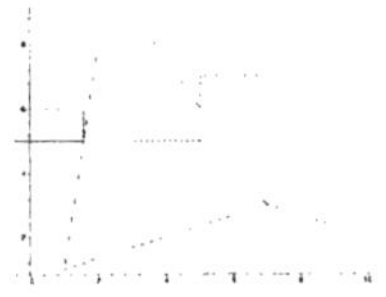

Figure 15: A final result by the cost-first method

Figure 16: Computed candidate places by the intersection-first method

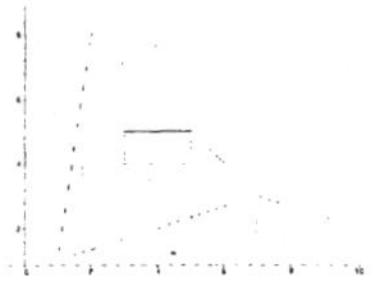

Figure 17: A final result by the intersection-first method

Next, the cost is determined by the following strategy: The closer a candidate place is to the middle point of the edge, the smaller its cost is set. Also, positions in the upper side of the edge is set to have less cost. See Fig.13.

We then apply a greedy approach which maintains a set of candidate places as processing edges one by one in some fixed order. In processing an edge, this set may be made smaller by deleting candidates in the following two manners:

Cost-first greedy approach In processing an edge e, for any intersecting pair of a candidate place for e and a candidate place for some other e', the place having the higher cost among two is removed from the set. Fig.14 is an example output. After processing all edges, we select the lowest cost candidate place among remaining candidate places for each edge. The result for Fig.14 is Fig.15.

Intersection-first greedy approach In this approach, each candidate place is associated further with the number, called the intersection number, of other candidate places, in the current set, intersecting with it. Then, the algorithm proceeds in a similar way by removing the candidate place having higher intersection number. Fig.16 is an example output, for the same data set in Fig.14. As in the case of the cost-first greedy approach, at the end, we select the least cost candidate place among remaining candidates for each edge. For Fig.16, a result becomes as in Fig.17.

Comparing Fig.14 and Fig.16, the final candidate set in Fig.14 is of size 16, while 18 in Fig.16. Concerning the final results, the intersection-first method is superior in this case. Also, it gives a well-balanced placement.

The above method can be extended to place a label to a polygonal line. A subway line may be represented by a polygonal line whose nodes are stations and edges are subway lines connecting adjacent stations. In such cases, just one placement of a subway line label suffices in the subway map. The above method can be extended to this case.

Fig.18 and Fig.19 are results for the Tokyo metropolitan subway map by the cost-first and intersection-first greedy algorithms, respectively. In this case, many lines

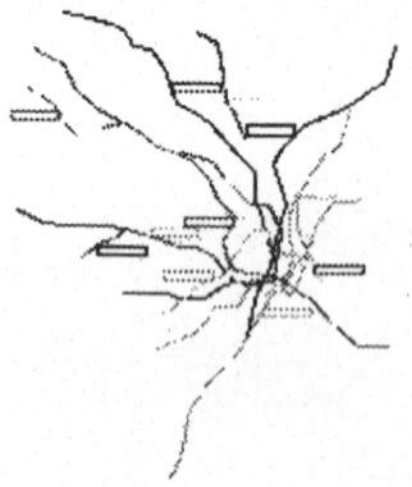

Figure 18: Cost-first method

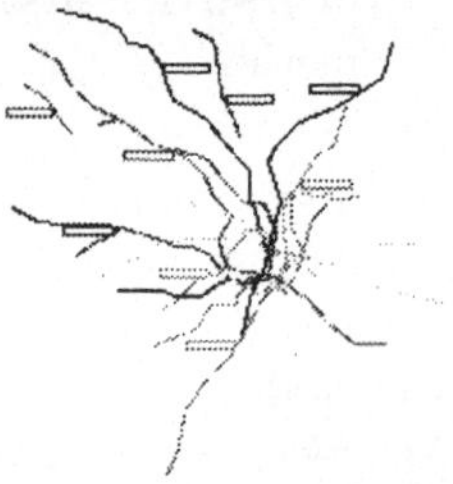

Figure 19: Intersection-first method

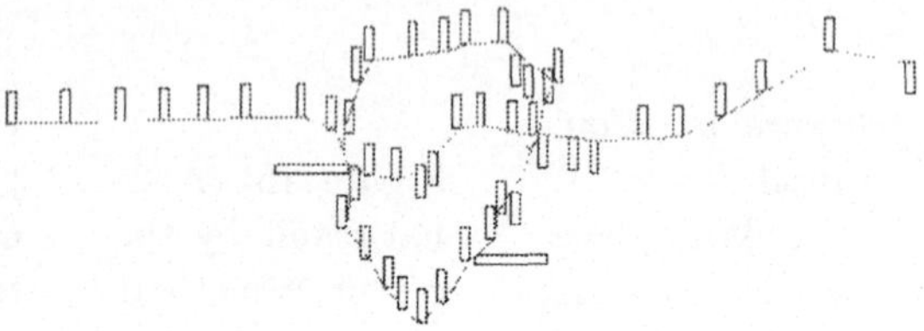

Figure 20: Station and line names of circular and central JR train lines

passes the central part of Tokyo, the cost of candidate places are slightly modified in such a way that one-third and two-third positions of a polygonal line has zero cost. In this case, the intersection-first greedy algorithm performs better.

So far, we have dealt with the ELP problem for edge labels. In concluding this subsection, we show a result for the GFLP problem for node and edge labels for the train map with line names and station names. For point labels, four natural candidate positions (NW, NE, SW, SE) of a point are considered. Since line names have more freedom, the point labeling is put higher priority. Fig.20 is a result for JR lines, circular line and central line, in Tokyo. For 47 out of 48 stations, together with two line labels.

4 Finding Detours in ITS

For the car navigation system, the most typical query is a shortest-path query. This query is very important in mobile computing environments based on GIS [5]. As dynamical traffic information newly becomes available such as ATIS and VICS mentioned in the introduction, more sophisticated queries come to be required. Also, the static geographical database of roads itself has grown up further, and similarly in this respect advanced types of queries are necessary to realize a user-friendly interface meeting the current circumstances. One important query among them is a detour query which provides information about detours; for example, enumerating several candidates for useful detours. We have proposed an efficient algorithm for enumerating meaningful detours [14].

'Detour' is not so clear concept. Thus we must define it precisely. The k-th shortest paths for moderate k have severe overlap with the shortest path in most cases, and are not suitable as good detours. Taking such overlaps into consideration, we define 'detour' as follows [14]:

Definition 1 *'Detour' is Δ longer than the shortest path at most, branch off and join the shortest path only once, and has the smallest overlap with the shortest path among such paths. If several paths satisfies these constraints, choose the shortest one.*

Figure 21: Explanation of the definition of 'detour'

In Fig.21, there are two 'branchings', and each may be used separately in our detours, but, the path using both of them is neglected in our approach, since it can be generated from the previous two paths.

In Shibuya et al. [14], an efficient network algorithm is given to find thus defined detours. We here only show computed examples by this algorithm. Fig.22 shows the obtained detours from Sayama to Matsudo, on an real road network database of Kanto district area in Japan, when Δ is 100 seconds and 120 seconds. In the figure, the thickest line is the shortest path, and the relatively thinner line which branch off it is the obtained detour.

(1) $\Delta = 100$

(2) $\Delta = 120$

Figure 22: Detours between Sayama and Matsudo

Acknowledgment

The work of the first, second and fourth authors was supported in part by the Grant-in-Aid for Scientific Research of the Ministry of Education, Science, Sports and Culture of Japan. The work of the second and fourth authors was supported in part by the High Technology Research Center Project, "Integrated Geographic Information Systems" of Chuo University.

Bibliography

1) H. Aonuma, H. Imai and Y. Kambayashi: A Visual System of Placing Characters Appropriately in Multimedia Map Databases. *Proceedings of the IFIP TC 2/WG 2.6 Working Conference on Visual Database Systems*, North-Holland, 1989, pp.525–546.

2) L. P. Chew: Constrained Delaunay Triangulations. *Algorithmica*, Vol.4 (1989), pp.97–108.

3) H. Edelsbrunner and T. S. Tan: An Upper Bound for Conforming Delaunay Triangulations. *Discrete and Computational Geometry*, Vol.10 (1993), pp. 197–213.

4) M. Ester, H.-P. Kriegel and X. Xu: Knowledge Discovery in Large Spatial Databases: Focusing Techniques for Efficient Class Identification. *Proceedings of the 4th International Symposium on Large Spatial Databases*, 1995, pp. 67–82.

5) T. Ikeda, M.-Y. Hsu, H. Imai, S. Nishimura, H. Shimoura, K. Tenmoku and K. Mitoh: A Fast Algorithm or Finding Better Routes by AI Search Techniques. *Proceedings of the International Conference on Vehicle Navigation & Information System (VNIS'94)*, 1994, pp.90–99.

6) H. Imai and M. Inaba: Geometric Clustering with Applications. *Zeitschrift für Angewandte Mathematik und Mechanik (ZAMM)*, Vol.76, Suppl. (1996), pp.183–186.

7) H. Imai and M. Inaba: Geometric Clustering by Divergence and Its Underlying Discrete Proximity Structures. *IEICE Trans. Information and Systems*, Special Section on Discovery Science, Vol.E83-D, No.1 (2000), pp.27–35.

8) K. Imai and T. Kameda: Label Placement Problem and Its Application to Train Line Maps (in Japanese). *Proceedings of the 5th Symposium on Integrated Geographic Information Systems*, Chuo University, 2000, pp.17–24.

9) M. Iri (director), T. Koshizuka (editor): *Computational Goemetry and Geographical Information Processing* (in Japanese). Kyoritsu-Shuppan, 1993.

10) M. Iri, K. Murota and T. Ohya: A Fast Voronoi-Diagram Algorithm with Applications to Geographical Optimization Problems. *Proceedings of the 11th IFIP Conference on System Modelling and Optimization*, Lecture Notes in Control and Information Science, Vol.59, Springer-Verlag, 1984, pp.273–288.

11) K. G. Kakoulis and I. G. Tollis: An Algorithm for Labeling Edges of Hierarchical Drawings. *Proceedings of Graph Drawing '97*, Lecture Notes in Computer Science, Vol.1353 (1998), pp.169–180.

12) K. G. Kakoulis and I. G. Tollis: A Unified Approach to Labeling Graphical Features. *Proceedings of the 14th ACM Symposium on Computational Geometry*, 1998, pp.347–356.

13) K. Kubota: A Numerical Method for Extracting Topological Information from Digital Maps. *Proceedings of the 2nd Symposium on Integrated Geographic Information Systems*, Chuo University, 1998, pp.63–68.

14) T. Shibuya, T. Ikeda, H. Imai, S. Nishimura, H. Shimoura, and K. Tenmoku: Finding a Realistic Detour by AI Search Techniques. *Proceedings of the 2nd Intelligent Transportation Systems (ITS'95)*, 1995, pp.2037–2044.

15) K. Sugihara: *FORTRAN Computational Geometry Programming* (in Japanese). Iwanami, 1998.

16) A. Wolff: Map Labeling. `http://www.math-inf.uni-greifswald.de/map-labeling/`.

10

Management and Multimedia Presentation of Heterogeneous Documents

Kazumasa Yokota
Takeo Kunishima
Akiyoshi Matono
Okayama Prefectural University

Bojiang Liu
Okayama University of Science

ABSTRACT

XML (eXtended Markup Language) has become almost de facto standard for structuring digital documents and been being extended to cover multimedia information by efforts of many languages such as SVG, X3D, and VoiceXML. However, such extensions are limited under the syntax of XML, which is not necessarily appropriate for users' semantics-oriented treatment such as representing relationships among objects and attaching memos to shared documents. In this article, we consider another approach to treat semantically extended management of XML-based documents to cope with various applications. As our approach focuses on not only semistructured data but also structured data such as complex objects, conventional query processing can be also applied to a set of documents. Further, by introducing logical elements, it is also useful for reducing design heterogeneity and content-based classification.

1 Introduction

With rapid advances of computers and networks, almost every information has been being digitalized. Among them, a concept of *digital document* plays a central role: that is, it does not necessarily correspond to the form of paper, but includes not only texts and pictures but also movies, programs, and even virtual worlds. In other words, digital documents are essentially multimedia-based and, therfore, their models and presentation methods have to be reconsidered, differently from conventional documents'.

XML (eXtensible Markup Language) has become almost *de facto* standard as a common protocol for exchange of digital documents, and its multimedia applications such as SVG, X3D, VoiceXML, and SMIL have been developed. However, as application areas are enlarging, we meet more restrictions of XML: for example, from a data model pont of views such as complex objects and semistructured data, from viewpoints of combination of shared and personal information, and from viewpoints of integration of multimedia information. It is indispensable to relax such restrictions for advanced applications with electronic documents.

Further, from a user point of view, multimedia documents should be interactive to users, because users want to read through a forest of multimedia information.

We are now engaged in heterogeneous document management [Kunishima et al 99)], digital theme parks [Liu et al 00)], interactive synchronous presentation of multimedia information[Fujino et al 00)], literature databases[Miyake et al 00)], and so on, where we require extensions of XML and interactive presentation of multimedia information, and have implemented their prototype systems.

In this article, we discuss relaxation of such restrictions, or semantic extensions of XML, and interactive multimedia presentation. We report our approaches to their problems.

First, we discuss heterogeneities in management of digital documents and requirements of extensions of XML in Section 2. Secondly, we discuss interactive presentation of multimedia documents, especially authoring facilities, in Section 5. Thirdly, we propose a model of semantic extensions of XML in Section 3, its architecture for interactive presentation in Section 5, and introduce an implementation of our prototype systems in Section 4. Lastly we summarize their technologies in Section 6.

2 Approachs to Management of Heterogeneous Documents

Consider that we write a paper in LaTeX. It is compiled into a dvi file. The dvi file is tranformed into a postscript file for printing and into a pdf file for online publication. We sometimes write an oral manuscript and prepare a PowerPoint file for presentation. Although their contents are the same, we have to produce various files.

There are various heterogeneities in digital documents:

- We can represent documents in various data format such as plain text, LaTeX, postscript, PDF, MS Word, MS PowerPonit, HTML, and XML. To manage their content-based relationships, we have to consider links between objects in various formats.

- As we want to write personal information even in a shared document, the documents must be viewed with different contents from each user. There may be some shared levels: not only personal but also group levels, and we can consider a shared hierarchy and view mechanisms according to some purposes.

- Even if the content is the same, it might be designed to be physically differently divided documents. In such cases, we must guarantee the same result for any query to their different designed documents with the same content.
- Although a document may be updated as different versions, we have to manage their versions under to one root, or multiple ones.
- As a digital document may have various multimedia information such as voice, picture, movie, program, and virtual world, their integrated treatment is indispensable.

As XML is almost *de facto* standard as document format and covers formats of many multimedia information, we consider that it is appropriate to employ XML as the common protocol of various data formats. The problems are how to cope with other heterogeneities, how to define logical objectsm which does not correspond to the syntax of XML, and how to treat conventional structured data embedded in XML documents.

2.1 Features of XML

Let me first consider main features of XML. A document structured in XML is considered as a tree, where each subtree corresponds to an element and its root node has a label called an element name or a tag. A leaf corresponds to PCDATA or a null value. Each node may has a set of attributes. From a database point of view, its characterics are as follows:

- *attribute (⟨l-attr⟩)*: It corresponds to a tuple as in a relation, where a label appears only once, a value is atomic, and the order of attributes is arbitrary.
- *element (⟨e-attr⟩)*: It can have multiple attributes with the same label, while their ordering is meaning and cannot be a set. Further, in elements, PCDATA might be mixed with other elements at the same level. Therfore the value of this subtree must be concatenated in a depth-first manner.
- *identifier*: In XML, *id*-attribute and *idref*-attribute are pre-defined, while their meaning is too weak and they cannot be used as an identifier.

The semantics of XML is not strictly defined and application designers can define additional semantics. Although the semantics of tags can be defined in DTD, it is not ordinary semantics but rather syntactical restrictions. As its expressive power is poor from a database point of view, additional semantics for data models are introduced in various proposals such as XML schema[XML Schema 01] and XML Query Data Model[XML Query Data Model 01]. In this article, we will consider another semantic extensions to treat heterogenenous multimedia documents: that is, introduction of identities, equality constraints, constructors into elements, dummy tags for PCDATA, and so on, not only for data models but also knowledge representation.

2.2 Requirements of Data Models

In documents, we consider that structures and semistructured data are mixed. Therfore conventional concept of complex objects with identities should be embedded in XML. Although a concept of semistructured data includes XML, XML is poor in data constructors. We require the followings:

- *Data constructors*: A record or a tuple is defined as a set of attributes (attribute-value pairs), whose ordering is arbitrary. So we have to introduce constructors to describe a tuple and a set.
- *Self-description*: As the semantics of data cannot be necessarily described in its scheme, additional semantics is written with data itself. Newly introduced special attributes in this article corresponds to this one: that is, depending on special attribute values, interpretation of an element can be changed.
- *Identities*: As in OEM[Abiteboul et al 99)], by attaching object identifiers to all elements, it is possible to describe links among objects, while by considering partial information of semistructured data, static identifiers might be not so better. We introduce both object identifiers for describing partial information.

Generally, although semistructured data includes structured one, the former does not necessarily guarantee not only the latter convenient representation but also its semantics. We have to introduce data constructors, identifiers, and their semantics.

2.3 Requirements of Knowledge Representation

Besides the above requirements in Section 2.2, we have to consider the followings:

- *Logical elements*: Although the structure of XML is a tree, objects are generally more complex, that is, a graph. Even in digital documents, we require to memoize arbitrary parts as objects, which might not follow the syntax of XML. To keep the syntax of XML, we must introduce a concept of logical objects, which might divide the logical element into necessary number of segments or group multiple XML objects, however we must guarantee the semantics of logical elements.
- *Equality constraint*: As identitifiers are used to indentify specified instance objects, they cannot be used to represent general circular structure and mutually related partial information. Although identities are generally global, equality constraints are locally more powerfull than identitities.
- *Links and navigation*: As more links are used to manage digital documents for reuse and maintenance than in other applications, it is desired that retrieval functions are declaratively defined, although navigations have been only procedurally defined in most systems.

As knowledge representation is generally complex, we focus mainly on knowledge representation aspects of digital documents as the above.

2.4 Other Requirements

As digitald documents are frequently shared and accessed through networks by many users, their contents have to be prepared to satisfy requirements of many users such as interests and levels For example, some user might want only a specific topic with explanation for novices and his friends' memos. In such cases, necessary contents shaould be dynamically extracted from the shared document. As such multifold information cannot be obtained by conventional information retrieval technologies, multifold information has to be embedded inside digital documents.

As digital documents are esssntially multimedia-based, we can apply this application to others such as digital theme parks. In such an application, we require additional extensions according to specific domains.

3 Extensions of XML for Multimedia Document

3.1 Graph Framework of XML

First, consider a simple example of XML in Figure 1. Although there are many

```
⟨book-order⟩
   ⟨customer⟩Yokota⟨/customer⟩
   ⟨shop location="Soja"⟩ABC⟨/shop⟩
   ⟨goods⟩
      ⟨book⟩
         ⟨publish year="1999"/⟩
         ⟨name⟩Quixote⟨/name⟩
      ⟨/book⟩
      ⟨book⟩
         ⟨publish year="2000"/⟩
         ⟨name⟩Dierdre⟨/name⟩
      ⟨/book⟩
   ⟨/goods⟩
⟨/book-order⟩
```

Figure 1: Example of XML

approches to formalization of XML, we consider an XML document as a graph with labeled edges or a set of graph equations. Figure 1 is represented as a graph with labeled edges as in Figure 2.

Figure 2 can be represented as a set of graph equations as in Figure 3, where "_" is an anonymous variable or a dummy value.

Although this definition might be redundant, it is very useful to define special attributes such as constructors, for example, into ⟨*goods*⟩ of the above example. In Abiteboul et al 99), ssd-expression for representing semi-structured data is defined.

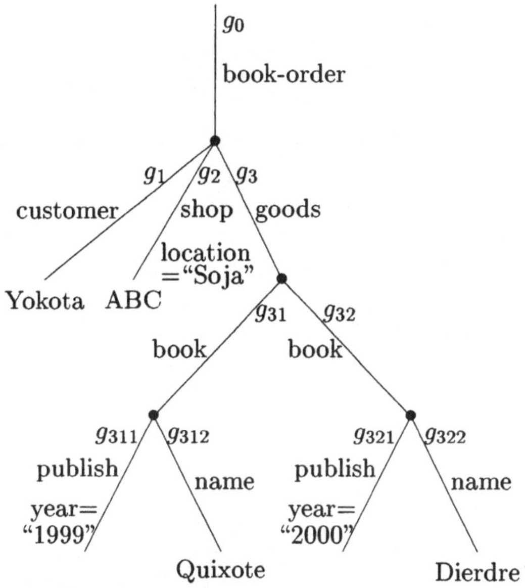

Figure 2: Graph Representation of XML

XML is generally defined as a graph model (for example, see [Abiteboul et al 99]). In this article, we define an XML object (element) as a set of graph equations, each of which is defined as follows;

$$
\begin{aligned}
\langle graph \rangle &= \{\langle equation \rangle, \cdots\} \\
\langle equation \rangle &= \langle g\text{-}variable \rangle = \\
&\quad (\langle id \rangle, \\
&\quad \langle e\text{-}label \rangle, \\
&\quad \{\langle a\text{-}label \rangle = \langle a\text{-}value \rangle, \cdots\}, \\
&\quad \{\langle e\text{-}value \rangle, \cdots\}) \\
\langle e\text{-}value \rangle &= \langle \text{PCDATA} \rangle\} \mid \langle \text{id} \rangle \mid \langle \text{g-variable} \rangle
\end{aligned}
$$

where $\langle g\text{-}variable \rangle$ is a variable for equality constraints among equations, $\langle id \rangle$ is an identifier of $\langle equation \rangle$, $\langle e\text{-}label \rangle$ is an element label, $\langle a\text{-}label \rangle = \langle a\text{-}value \rangle$ is an attribute of $\langle e\text{-}label \rangle$, and $\{\langle e\text{-}value \rangle, \cdots\}$ is an element value of $\langle e\text{-}label \rangle$. Each $\langle e\text{-}value \rangle$, called a sub-element of $\langle g\text{-}variable \rangle$ in the left hand of "=", and is defined as $\langle g\text{-}variable \rangle$, $\langle \text{PCDATA} \rangle$, or $\langle id \rangle$. $\langle \text{PCDATA} \rangle$ is an atomic value.

There are differences between an identifier and an equality constraint. Graphs with the same $\langle id \rangle$ have to be merged due to the property of identity. For merging graphs, subgraphs with different $\langle id \rangle$s cannot be merged, while subgraphs with different $\langle graph \rangle$ can be merged by unification.

$$
\begin{aligned}
g_0 &= (\_, \textit{book-order}, \_, \{g_1, g_2, g_3\}) \\
g_1 &= (\_, \textit{customer}, \_, \{\text{Yokota}\}) \\
g_2 &= (\_, \textit{shop}, \{\textit{locator} = \text{“}\textit{Soja}\text{”}\}, \{\text{ABC}\}) \\
g_3 &= (\_, \textit{goods}, \_, \{g_{31}, g_{32}\}) \\
g_{31} &= (\_, \textit{book}, \_, \{g_{311}, g_{312}\}) \\
g_{311} &= (\_, \textit{publish}, \{\textit{year} = \text{“1999”}\}, \_) \\
g_{312} &= (\_, \textit{name}, \_, \{\text{Quixote}\}) \\
g_{32} &= (\_, \textit{book}, \_, \{g_{321}, g_{322}\}) \\
g_{321} &= (\_, \textit{publish}, \{\textit{year} = \text{“2000”}\}, \_) \\
g_{322} &= (\_, \textit{name}, \_, \{\text{Dierdre}\})
\end{aligned}
$$

Figure 3: Graph Equations of XML

Two kinds of attributes correspond to XML's: we call them label attributes (abbreviated by $\langle l\text{-}attr\rangle$) and element attributes (abbreviated by $\langle e\text{-}attr\rangle$), respectively.

This syntax follows $\mathcal{QUIXOTE}$'s one[Yasukawa et al 90)] and the semantics is also defined on hypersets[Aczel 88)] as in $\mathcal{QUIXOTE}$. Consider the following system of graph equations:

$$
\begin{aligned}
g_2 &= (\_, l, \_, \{g_4\}) \\
g_3 &= (\_, l, \_, \{g_5\})
\end{aligned}
$$

If each of graphs does not have an identifier and any label attributes, they can be reduced by giving $g_2 = g_3$ as follows:

$$g_1 = (\_, l, \_, \{g_4, g_5\})$$

As for graph isomorphism, we require additional rules.

Consider an example which means a person deelpy in love with each other::

$$X : person[loves \Rightarrow person[loves \Rightarrow X]]$$

This is a circular one. In our extended XML, it is written as follows:

$$
\begin{aligned}
&\langle persone\_cntr = X\rangle \\
&\qquad \langle loves\rangle Y \langle /loves\rangle \\
&\langle /preson\rangle \\
&\langle persone_c ntr = Y\rangle \\
&\qquad \langle loves\rangle X \langle /loves\rangle \\
&\langle /person\rangle
\end{aligned}
$$

The graph representation is as follows:

$$\begin{aligned} X &= (\_, person, \_, \{g_1\}) \\ g_1 &= (\_, loves, \_, \{Y\}) \\ Y &= (\_, person, \_, \{g_3\}) \\ g_3 &= (\_, loves, \_, \{X\}) \end{aligned}$$

Consider another example.

$$\begin{aligned} & person[user\_id = id_{Kazu}, \\ & \qquad name = Yokota, \\ & \qquad spouse = id_{Yuko}, \\ & \qquad hobby = \{golf, tennis\} \\ & \qquad] \\ & Person[user\_id = id_{Yuko}, \\ & \qquad spouse = id_{Kazu}, \\ & \qquad \ldots] \end{aligned}$$

It is an example of complex object with identifiers. It is written as a graph as follows:

$$\begin{aligned} g_0 &= (id_{Kazu}, person, \{const = \text{“}set\text{”}\}, \{g_1, g_2, g_3\}) \\ g_1 &= (\_, name, \_, \{Yokota\}) \\ g_2 &= (\_, spouse, \_, \{id_{Yuko}\}) \\ g_3 &= (\_, hobby, \{const = \text{“}set\text{”}\}, \{g_{31}, g_{32}\}) \\ g_{31} &= (\_, pc\_data, \_, \{golf\}) \\ g_{32} &= (\_, pc\_data, \_, \{tennis\}) \\ g_4 &= (id_{Yuko}, person, \{const = \text{“}set\text{”}\}, \{g_5, g_6, g_7\}) \end{aligned}$$

In this article, from an application point of view, we will discuss rather XML-related issues than its formal semantics, due to restriction of space.

Our model includes this expression. In the rest part of this section, we consider requirements from an application point of view and, in the next section, discuss necessary special attributes.

3.2 Proposed Model

This proposed model, defined in Section 3.1, is a hypergraph structure with references. Compared with (complex) object models, there are the following differences:

- As for reference relations, there are two kinds of concepts: identifier $\langle id \rangle$ and equality constraint $\langle g\text{-}variable \rangle$, both of which can be used if necessary.

- There are two kinds of attributes, ⟨*l-attr*⟩ and ⟨*e-attr*⟩: the former is based on a tuple constructor and the latter is based on a set (tuple) constructor or a list constructor, specified in ⟨*l-attr*⟩. A set or a tuple can be differentiated by differences of elements.

Although ⟨*l-attr*⟩ seems to be similar with F-logic's label's attributes[Kifer et al 89)], ⟨*l-attr*⟩ is in more general form.

3.2.1 Introduction of Specific Attributes

To extend XML semantically, we introduce special attributes and extend the meaning of element attributes as follows:

- *element_id* (*e_id*) = "xxx": specification of the identifier, "xxx", of the element which includes this element
- *e_constraint* (*e_cons*) = "*variable*": specification of the variable, "*variable*", as equality constraint which includes this element
- *constructor* (*const*) = "*set*" | "*tuple*" | "*list*" | "<u>*sequence*</u> (*seq*)": specification of a constructor of the following sub-elements
- *doc_group* = "*yes*" | "<u>*no*</u>": specification of a document group
- *user_name* (*u_name*) : xxx = "*yes*": specification of an owner of this element
- *parent* = "*element_id*": specification of the parent element, "*element_id*", which includes this element
- *condition* (*cond*) : *nnn cond* : xxx = "*yes*"|"*no*": specification of a condition for activating the following elements, where *nnn* is an selective labelsand xxx is its corresponding value
- *course* = "<u>*both*</u>" | "*one*": specification of ordering of sub-elements

where an underline means a default value.

Identifiers specified by *e_id* may be used as attribute values or element values. Elements with the sama *e_id* or *e_cons* must be the same. Is they are not same, they are unified. Elements with the same *eid* may have diffrent *e_cons*, while elements with the same *e_cons* may not have different *e_id*.

doc_g guarantees that the folowing element consists one logical unit by query transformation.

As *user_name* and *cond* might overlap elements and do not judge the structure automatically, *parent* specifies the unique structure. If a *condition* is satisfied for a given parameters, the following sub-elements are activated. *course* is a special label for directions among sub-elements as in a tour course in digital theme parks.

Before considering data operations, we have to make various concepts clear: relation between graph and its corresponding value, relation between native and logical elements, and so on.

3.2.2 Combination of PCDATA and Text

First, we introduce a special tag $\langle pc\_data \rangle$ to treat text data uniformly. Consider a data with PCDATA,

$$\cdots \langle a \rangle T_1 \langle b \rangle T_2 \langle /b \rangle T_3 \langle /a \rangle \cdots$$

where T_1 and T_3 are PCDATA, we describe as follows:

$$\begin{array}{l} \cdots \langle a \rangle \langle pc\_data \rangle T_1 \langle /pc\_data \rangle \\ \quad \langle b \rangle T_2 \langle /b \rangle \langle pc\_data \rangle T_3 \langle /pc\_data \rangle \langle /a \rangle \cdots \end{array}$$

By such description, we can avoid to mix tagged sub-elements with naked PCDATA.

The treatment of text is different depending on an attribute *const*. Let *text* and *cat* be functions of text extraction and string concatenation, respectively. *text* is recursively defined as follows:

$$\begin{array}{l} text(T) = T \qquad \text{where } T \text{ is text} \\ text((\_, pc\_data, \_, \{T\})) = T \\ text((\_, l, \{const = \text{“}seq\text{”}\}, \{a_1, a_2, \cdots, a_n\})) \\ \quad = cat([text(a_1), text(a_2), \cdots, text(a_n)]) \\ text((\_, l, \{const = \text{“}set\text{”}\}, \{a_1, a_2, \cdots, a_n\})) \\ \quad = \{text(a_1), text(a_2), \cdots, text(a_n)\} \\ text((\_, l, \{const = \text{“}list\text{”}\}, \{a_1, a_2, \cdots, a_n\})) \\ \quad = [text(a_1), text(a_2), \cdots, text(a_n)] \end{array}$$

In a case of “*list*”, text is concatenated in a depth-first mode, while in a case of “*set*”, elements are not concatenated and duplicated elements are reduced: *text* is nested when “*set*” constructors are nested. The treatment of *tuple* is the same with “*set*”. In a case of “*seq*”, text is concatenated in depth-first and left-to-right manner.

This function, *text*, is use to read or extract text from a given graph. Note that a set of values is extracted from a graph which semantically means a tuple. It corresponds to read data without label specification.

3.2.3 Logical Elements

For a description

$$\cdots \langle a \rangle T_1 \langle /a \rangle \langle b \rangle T_2 \langle /b \rangle \cdots$$

a logical element, $\langle c \rangle$, is generated when tags are overlapped as follows:

$$\cdots \langle a \rangle T_{11} \underline{\langle c \rangle T_{12} \langle /a \rangle \langle b \rangle T_{21} \langle /c \rangle} T_{22} \langle /b \rangle \cdots$$

Considering the well-formedness of XML, the above text should be segmented as follows:

$$\langle p \rangle \cdots \langle a \rangle T_{11} \underline{\langle c \rangle T_{12} \langle /c \rangle} \langle /a \rangle \langle b \rangle \underline{\langle c \rangle T_{21} \langle /c \rangle} T_{22} \langle /b \rangle \cdots \langle /p \rangle$$

However two parts with a tag $\langle c \rangle$, must be logically concatenated.

By introducing equality constraints (or identifiers) and a special attribute *parent*, we define it as follows:

$$g_0 = (\_, \_, \{const = \text{“}list\text{”}\}, \{g_1, g_2, g_3\})$$
$$g_1 = (\_, a, \{const = \text{“}list\text{”}\}, \{g_{11}, g_{12}\})$$
$$g_2 = (\_, b, \{const = \text{“}list\text{”}\}, \{g_{21}, g_{22}\})$$
$$g_3 = (\_, c, \{const = \text{“}list\text{”}, parent = \text{“}p\text{”}\}, \{g_{12}, g_{21}\})$$
$$g_{11} = (id_1, pc\_data, \_, \{T_{11}\})$$
$$g_{12} = (id_2, pc\_data, \_, \{T_{12}\})$$
$$g_{21} = (id_3, pc\_data, \_, \{T_{21}\})$$
$$g_{22} = (id_4, pc\_data, \_, \{T_{22}\})$$

where we can guarantee the ordering of text and avoid their duplication. From a logical element point of view, $\langle a \rangle$, $\langle b \rangle$, and $\langle c \rangle$ are logically equivalent, while an attribute *parent* is written in an attribute. Although the location is arbitrary, the semantics does not depend on the location but is the same. *text* in Section 3.2.2 can be applied also to this logical elements.

A similar concept is proposed for structuring genetic information [Stokes et al 99)]. As genetic information has cyclic and overlapping structure, XML is extended for the application. The point of this approach is to guarantee the consistency by considering the absolute address of genetic sequences. Our approach is based on identifiers instead of the absolute address.

3.2.4 Conditional Elements

By giving parameters (conditions) to an existing XML object, we want to obtain a new XML object, all of whose elements satisfy the conditions. For example, there are many seasonal elements in a digital document. In winter, we would like to walk through the course only with winter elements and unchangeable ones. Further, beginners want to read introductory explanation, while intermediaters want to skip redundant explanation. The following is an example of conditional elements:

$$\langle l \rangle \cdots \langle season\ cond : spring = \text{"}yes\text{"} \rangle \langle pc\_data \rangle T_1 \langle /pc\_data \rangle$$
$$\langle season\ cond : summer = \text{"}yes\text{"} \rangle \langle pc\_data \rangle T_2 \langle /pc\_data \rangle \langle /season \rangle \langle /season \rangle$$
$$\langle season\ cond : summer = \text{"}yes\text{"} \rangle \langle pc\_data \rangle T_3 \langle pc\_data \rangle \langle /season \rangle \cdots \langle /l \rangle$$

It is written as a graph as follows:

$$\begin{aligned}
g_l &= (_, l, _, \{\cdots, g_a, g_b, \cdots\}), \\
g_a &= (_, season, cond : spring = \text{“}yes\text{”}, \{g_{T_1}, g_{T_2}\}), \\
g_b &= (_, season, cond : summer = \text{“}yes\text{”}, \{g_{T_2}, g_{T_3}\}), \\
g_{T_1} &= (_, pc_data, _, \{T_1\}), \\
g_{T_2} &= (_, pc_data, _, \{T_2\}), \\
g_{T_3} &= (_, pc_data, _, \{T_3\})\ \}
\end{aligned}$$

We show another example:

$$\begin{array}{l}
\langle season\,cond : summer = \text{"}yes\text{"}\, cond : spring = \text{"}yes\text{"}\rangle \\
\qquad\qquad \langle flower\rangle \\
\qquad\qquad \langle season\,cond : spring = \text{"}yes\text{"}\rangle\langle li\rangle cherry\langle /li\rangle\langle /season\rangle \\
\qquad\qquad \langle season\,cond : summer = \text{"}yes\text{"}\rangle\langle li\rangle sunflower\langle /li\rangle\langle /season\rangle \\
\qquad\qquad \langle li\rangle rose\langle /li\rangle \\
\qquad\qquad \langle /flower\rangle \\
\qquad \langle explanation\rangle \\
\qquad \cdots\langle level\,cond : novice = \text{"}yes\text{"}\rangle\cdots\langle /level\rangle\cdots \\
\qquad \cdots\langle season\,cond : spring = \text{"}yes\text{"}\rangle\cdots\langle /season\rangle\cdots \\
\qquad \langle /explanation\rangle \\
\langle /season\rangle
\end{array}$$

We can write simple conditions in $\langle$*l-attr*$\rangle$, however we have to use $\langle$*e-attr*$\rangle$ or name space for complex conditions, because $\langle$*l-attr*$\rangle$ is only in the form of a flat tuple.

3.2.5 Document Groups

A document group is a logical unit of multiple documents to be treat as one. The group is defined as follows:

$$\begin{aligned}
g &= (_, doc, \{doc_group = \text{"}yes\text{"}\}, \{g_1, g_2, \cdots, g_n\}) \\
g_i &= (_, chapter, _, \{d_{i1}, \cdots\})
\end{aligned}$$

where g is a logical document consisting of g_1, g_2, $\cdots$, g_n. As this unit is arbitrary, it must be explicitly specified. If the ordering among elements is important, $const =$ "*list*" can be specified with *doc_group*.

4 Document Operations

To define document operations, we have to make various concepts clear: relation between graph and access path, logical operations, and normalization of a set of graphs or access paths.

4.1 Access by Paths and Identifiers

We can consider two different accesses as follows:

$$\begin{array}{rcl}
\langle a\text{-}path\rangle & ::= & \langle graph\rangle[.\langle path\rangle] \\
& | & \langle graph\rangle[.\langle path\rangle].\langle a\text{-}label\rangle \\
\langle path\rangle & ::= & [\langle path\rangle.]\langle e\text{-}label\rangle \\
& | & [\langle path\rangle.]list(\langle num\rangle)
\end{array}$$

where $list(\langle num \rangle)$ specifies an n-th element in a list. Although the ordering in a set is meaningless, it is convenient to specify $set(\langle num \rangle)$ for taking a specific element. We define $set(\langle num \rangle)$ as an n-th element arranged by a system.

For access paths a_1, a_2, we define $a_1 \preceq a_2$ when there exists $\langle path \rangle$ p and $a_1.p = a_2$.

By applying $\langle a\text{-}path \rangle$ for a given graph, we can specify a substructure of a graph as follows:

$$\begin{aligned} g &= (id, el, \{al_1 = av_1, \cdots\}, \{g_1, g_2, \cdots, g_n\}) \\ g_i &= (id_i, el_i, \_, \{g_{i1}, \cdots\}) \end{aligned}$$

In this case, we define as follows:

$$\begin{aligned} g.el &= \{g_1, g_2, \cdots, g_n\} \\ g.al_i &= av_i \mid \top \\ g.el.el_i &= \{g_{i1}, \cdots\} \mid \top \\ g.el.list(i) &= \{g_i\} \mid \top \qquad \text{in a case of } const = \text{“}list\text{”} \end{aligned}$$

We can get the same result by using identifiers id instead of graphs. It can be extended to a set of graphs as follows:

$$\begin{aligned} \{g_1, \cdots\}.al &= \{g_1.al, \cdots\} \\ \{g_1, \cdots\}.el &= \bigcup_i (g_i.el) \end{aligned}$$

We call a graph with a sequence of labels an access path. Here $\top$ is a value, when a specified label does not exist. In a case of $const$ = “$list$”, $\top$ cannot be omitted, while in a case of $const$ = “set”, $\top$ and same or redundant elements are reduced.

For example, g_0.*book-order.goods.book* in Figure 3 points $\{g_{312}, g_{322}\}$.

4.2 Basic Retrieval

A query is a logical combination of unit queries, each of which is in the following form:

$$\langle a\text{-}path \rangle \ \langle comparison\ operator \rangle \ \langle search\ condition \rangle$$

Retrieval is executed for a set of $\langle l\text{-}attr \rangle$ or $\langle e\text{-}attr \rangle$ specified by access paths. Data is generated by the definition in Section 3.2.2. Data included in logical elements is similarly generated. The result of a unit query is a set of (sub)graphs. For example, in Figure 3, a query, g_0.*book-order.goods.book.publish* = 2000 returns $\{g_{321}\}$.

Remark that a graph can be transformed into a set of access paths. For example, g_{321} in the above example can be transformed into a set of an access path, g_0.*book-order.goods.book*$[set(2)]$ by adding $set(\langle num \rangle)$ or $list(\langle num \rangle)$ for identifying an element of a set to its corresponding label. If g_{321} is referred by multiple nodes, we have to define a domain where access paths are effective because we can get multiple access paths.

In querying to a graph, regular expressions are usually used as specification of an access path (for example, Lorel, UnQL, XML-QL) in [Abiteboul et al 99]). Such expressions are developed into a set of possible access paths and multiple queries can be generated. It is important for users, however we omit it in this article, because it is not essential.

4.3 Logical Operations and Reduction of Access Paths

We define logical operations between graphs obtained in Section 4.2. Let $a\text{-}path(g)$ be a set of access paths, obtained by transformation of a graph g. Transformation of $\{g_1, g_2, \cdots\}$ results in $\bigcup_i(a\text{-}path(g_i))$. It requires to specify a domain where reference relations are effective, as already mentioned.

Given a set of graphs or access paths, we have to normalized it, that is, eliminate duplication and redundancy. As for graphs, we explain in Section 4.4.

As, when $a_1 \preceq a_2$, a_2 has more specific information than a_1, we define Smyth ordering between sets of access paths S_1, S_2 as follows:

$$S_1 \preceq_S S_2 \stackrel{\text{def}}{=} \forall a_2 \in S_2, \exists a_1 \in S_1 . a_1 \preceq a_2$$

Let the minimum set be the representative of corresponding equivalence classes. A set of access paths as the result of retrieval is the representative in this meaning. We consider only such sets.

When $a_1 \preceq a_2$, we define $a_1 \wedge a_2 = a_1$, $a_1 \vee a_2 = a_2$. Further we define logical operations between sets of access paths as follows:

$$\begin{aligned} S_1 \wedge S_2 &= \{a_1 \wedge a_2 \mid a_1 \in S_1, a_2 \in S_2\} \\ S_1 \vee S_2 &= \{a_1 \vee a_2 \mid a_1 \in S_1, a_2 \in S_2\} \end{aligned}$$

As an access path specifies a subgraph, we can enlarge the graph by reducing the access path as follows:

$$\pi_{l_i}(g.l_1.l_2.\cdots.l_i.\cdots.l_n) = g.l_1.l_2.\cdots.l_i$$

which is similar with projection as in relational algebra. When the specified label appears in multiple times, we take the longest path: that is, $\pi_{l_i}(\pi_{l_i}(P))$ is not necessarily the same as $\pi_{l_i}(P)$. The result is normalized in the sense of Smyth ordering.

4.4 Merge of Graphs

In distributed environments, the same object might be located in different locations in different expressions. When we get such objects as a result of search operations, graphs should be normalized according to the semantics and graphs with the same identifier should be merged. Merging based on equality constraints is executed when new variable constraints are inserted or a user inserts such constraints. When identifiers and equality constraints are used, their effective domains become important, however we assume it global for simplicity.

As in the definition of a graph, an identifier is defined for one label. Let *id* and *lab* be functions to extract an identifier and a label from a given graph, for graphs g_1, g_2, $id(g_1) = id(g_2)$ and $lab(g_1) \neq lab(g_2)$ are inconsistent. That is, the definitions of g_1, g_2 become dummy ($\perp$). In other cases, each attribute should be merged as follows:

- In a case of $\langle$*l-attr*$\rangle$, a same label must have the same value, otherwise inconsistent. $\langle$*l-attr*$\rangle$'s with different labes are merged.
- In a case of $\langle$*e-attr*$\rangle$, a constructor should be same. In a case of set, merging is required, while, in a case of list, concatenation is required.

For example, when

$$g_1 = (id, el, \{const = \text{“}set\text{”}\}, \{g_{11}, g_{12}\})$$
$$g_2 = (id, el, \{const = \text{“}set\text{”}, shop = \text{“}Soja\text{”}\}, \{g_{21}, g_{22}\})$$

g_1 and g_2 is merged into

$$g_3 \quad = \quad (id, el, \{const = \text{“}set\text{”}, shop = \text{“}Soja\text{”}\}, \{g_{11}, g_{12}, g_{21}, g_{22}\})$$

Its result, $\langle$*e-attr*$\rangle$, is also merged if it satisfies the above conditions. we can consider to enhance the possibility of merging by introducing type concepts as in ψ-term[Ait-Kaci 84)].

In a case of equality constraints, when

$$g_1 = (\_, el, \{const = \text{“}set\text{”}\}, \{g_{11}, \cdots\})$$
$$g_2 = (\_, el, \{const = \text{“}set\text{”}\}, \{g_{21}, \cdots\})$$

the following merge is executed

$$g_1 = g_2 = (\_, el, \{const = \text{“}set\text{”}\}, \{g_{11}, \cdots, g_{21}, \cdots\})$$

for a given constraint $g_1 = g_2$. This operation is required when objects which had be considered to be different are identified, or an equation is given by constraint propagation. We can get maximum information by taking minimum graphs.

4.5 Query Transformation for Document Groups

A document group guarantees that a query returns the same result, even if they consist of different documents. For the purpose, it is necessary to transform a query to a set of queries and a summation operation, depending on physical configuration[Kunishima et al 9
Consider a query $\sigma_F(g)$ (selection) for a document group in Section 3.2.5. It is transformed into $\alpha_{F_1}\sigma_{F_2}\{g_1, g_2, \cdots\} = \alpha_{F_1}\{\sigma_{F_2}(g_1), \sigma_{F_2}(g_2), \cdots\}$, where α_{F_1} is a summation operation.

For example, consider $F = a \wedge b$ for a document group $\{d_1, d_2\}$. A query $\sigma_{a\wedge b}$ is transformed $\sigma_{a\vee b}$ for d_1 and d_2, and, after the processing, its corresponding summation operation, $\alpha_{a\wedge b}$, checks whether d_1 satisfies $a \wedge b$, d_2 satisfies $a \wedge b$, d_1 satisfies a and d_2 satisfies b, or d_1 satisfies b and d_2 satisfies a.

4.6 Navigation

To link two graphs, we consider identifiers and equality constraints, which are different from *id* attributes and XPointer/XLink in XML. As they are essentially bi-directional, we consider only bi-directional links, even if they are defined to be only one-directional. For example, references are usually one-directional, while users want to use them as bi-directional links for retrieval.

Transformation from graphs to access paths, described in Section 4.2, corresponds to navigation via links. For example, consider a query, "Search papers which satisfy a condition C_1 and refer papers which satisfy a condition C_2," which can be divided into three queries: 1) "search papers which satisfy a condition C_2," and 2) "search papers which satisfy a condition C_1," and 3) "search papers contained in the results of 2) which refer any of the results of 1)." The results of 1) (a set of graphs) is transformed into a set of access paths in a domain including the results of 2).

5 Approachs to Interactive Multimedia Presentation

There are many systems for synchronizing multimedia information and presenting them: SMIL[SMIL 98)], TVML[TVML 99)], HyTime[HyTime)], and IMD[Vazirgiannis 99)]. However, corresponding to the above requirements of documents, we consider the following requirements:

- As the common protocol is XML, control languages should be compatible with XML.
- If users want to insert their own information into a common document during its presentation, the presentation should be interactive: that is, it had better combine with some authoring functions.
- As we cannot design a comprehensive control language for any purpose, we had better consider a metalanguage which can embed other languages into one presentation.

The second and third points require some programming features: that is, the second one corresponds to some debugging functions and the third one corresponds to meta-control functions. Most of conventional languages (or scripts) specify temporal and spatial synchronization and generate streaming data, and thus lack the above features.

In Section 3, we explain semantical (not syntactical) extensions of XML, but do not propose a new language. The reason is that we want to use various XML software environments. However, to control XML-based documents, it is indispensable to use a language which covers extensions of XML, specify temporal and spatial synchronization of multimedia information, call other presentation languages, and make users interact with presentation.

To control synchronization and interaction of multimedia information, we emply a knowledge representation language, QUIK[Liu et al 99)], as a control language of multimedia information. The language is an ancestor of $\mathcal{QUIXOTE}$[Yokota et al 99)], a DOOD language of the Japanese FGCS project. The reasons why we employ the language are following:

- As it is a powerful knowledge representation language, it is easy to represent semantical extensions of XML.
- The language supports representation and synchronization of multimedia information and it is easy to call other languages.
- As the language is also a programming language, it supports debugging functions: that is, it is possible to support interactive multimedia presentation functions such as stop, backward, forward, and change of temporal and spatial layouts of multimedia objects.

An XML document is processed as follows:

1. An XML document is pre-edited: that is, multimedia information is inserted into the document if necessary.
2. An XML document is converted into a QUIK program.
3. The QUIK interpreter executes a document in QUIK and presents multimedia information to a user.
4. A user can interrupt (stop) the execution at any time, see a synchronization graph, modify temporal and spatial layouts of multimedia objects, and re-start from arbitrary point.
5. A modified program can be strored as an XML document, is necessary, as a new version.

It is unnecessary for users to know a QUIK program. As for kinds of multimedia objects and synchronization mechanism, please refer [Fujino et al 00)].

6 Conclusions

We can see many enhancement requests of XML from viewpoints of many applications. In this article, we describe an extended model of XML from our applications' requirement point of view and their interactive presentation mechanism. XML is not syntactically extended, but semantically extended by introducing special attributes. The reason why we take this approach is that we keep upper compatibility with

XML as de facto standard and we use extended features. The semantics of extended features must be formalized, because it becomes too complex to follow the procedures as in the original XML. We had better add one advantage of our approach: that is, in this approach, it is rather easy also to introducing other additional features. However we have to introduce a language, QUIK, to represent and synchronize multimedia objects, and make interactive presentation possible.

We summarize the features of our model proposed in this article:

- Management of Heterogeneous Documents
  - We can use various constructors as in conventional data models.
  - We can define reference relations by identifiers and equality constraints, especially the latter of which is useful for describing partial information.
  - We can describe logical elements, which might be against the syntax of XML.
  - We can specify a set of documents with a content, naturally as a logical unit.
  - We can use navigation though links declaratively as transformation of subgraphs to access paths.
  - We can treat both shared and personal information uniformly.
  - It is easy to extend and implement our extensions.
- Interactive presentation
  - We can specify temporal and spatial layouts of multimedia information.
  - We can interpret an XML document as a QUIK program and present multimedia information as a result of the interpretation.
  - A user interrupts the execution (or presentation) at any time, modify multimedia objects and their layouts, and restart the presentation from any point.
  - We can store a QUIK program as an XML document in a database.

As future works, we consider the followings:

- Details of the formal semantics based on hypersets
- Implementation of extended XML
- Implementation of viewer (as a plug-in) and authoring tool
- Introduction of type information
- Design of a user-oriented query language
- Applications
  - Literature, documents, theme parks, and digital museum
  - Consistency with a DOOD language QUIK
  - Implementation of interactive drama performance and direction

Bibliography

Abiteboul et al 99) Serge Abiteboul, Peter Buneman, and Dan Suciu, *Data on the Web — From Relations to Semistructured Data and XML*, Morgan Kaufmann, 1999.

Aczel 88) P. Aczel, *Non-Well Founded Set Theory*, CSLI Lecture notes No. 14, 1988.

Ait-Kaci 84) Hassan Aït-Kaci, *A Lattice Theoretic Approach to Computation Based on a Calculus of Partially Ordered Type Structures*, Dissertation, Univ. of Pennsylvania, 1984.

Fujino et al 00) Takeshi Fujino, Issei Nomiya, Kazumasa Yokota, Takeo Kunishima, and Tadaaki Miyake, "Implementation of Interactive Drama Presentation System Based on Structured Documents," IPSJ SIGDBS, May, 2000. (in Japanese)

Kifer et al 89) M. Kifer and G. Lausen, "F-Logic — A Higher Order Language for Reasoning about Objects, Inheritance, and Schema", *Proc. ACM SIGMOD Int. Conf. on Management of Data*, pp.134-146, Portland, June, 1989.

Kunishima et al 99) Takeo Kunishima, Kazumasa Yokota, Bojiang Liu, and Tadaaki Miyake: "Towards Integrated Management of Heterogeneous Documents," *Cooperative Databases and Applications '99*, pp.39-51, Springer, Sep., 1999. available from http://alpha.c.oka-pu.ac.jp/ yokota/paper/codas99.ps

HyTime) Papers on HyTime, http://www.hytime.org/papers/

Liu et al 00) Bojiang Liu, Kazumasa Yokota, and Tatsuo Okamoto, "Considerations on Modeling Digital Theme Parks," Proc. IEICE Data Engineering Workshop, Mar., 2000. (in Japanese)

Liu et al 99) Bojiang Liu, Kazumasa Yokota, and Nobutaka Ogata, "Specific Features of the QUIK Mediator System," *IEICE Transaction on Information and System*, Vol. E82-D, No.1, pp.180-188, 1999. available from http://alpha.c.oka-pu.ac.jp/ yokota/paper/liu.ps

Miyake et al 00) Tadaaki Miyake and Kazumasa Yokota, "Literature Analysis as Information Processing — From Studies of the Dierdre Legend," Eigoseinen, vol.146, no.1, pp.6-9, Apr., 2000. (in Japanese)

SMIL 98) Synchronized Multimedia Integration Language (SMIL) 1.0 Specification, W3C Recommendation 15-June-1998, http://www.w3.org/TR/REC-smil/

Stokes et al 99) Aaron J. Stokes, Hideo Matsuda, and Akihiro Hashimoto, "GXML: A Novel Method for Exchanging and Querying Complete Genomes by Representing them as Structured Documents," IPSJ Trans. on Database, vol.40, no.SIG 6, pp.66-78, 1999.

TVML 99) TVML Language Specifications, http://www.strl.nhk.or.jp/TVML/English/E03.html.

Vazirgiannis 99) Michalis Vazirgiannis, *Interactive Multimedia Documents — Modeling, Authoring, and Implementation Experiences*, Springer LNCS 1564, 1999.

XML Query Data Model 01) XML Query Data Model, W3C Working Draft 15 February 2001, http://www.w3.org/TR/query-datamodel/.

XML Schema 01) XML Schema Part 0: Primer, W3C Recommendation 2 May 2001, http://www.w3.org/TR/xmlschema-0/; XML Schema Part 1: Structures, W3C Recommendation 2 May 2001, http://www.w3.org/TR/xmlschema-1/; XML Schema Part 2: Datatypes, W3C Recommendation 02 May 2001, http://www.w3.org/TR/xmlschema-2/.

Yasukawa et al 90) Hideki Yasukawa and Kazumasa Yokota, "Labeled Graphs as Semantics of Objects," IPSJ SIGDBS&SIGAI, Nov., 1990. available from http://alpha.c.oka-pu.ac.jp/ yokota/paper/sigya4.ps.

Yokota et al 99) Kazumasa Yokota, Hiroshi Tsuda, and Yukihiro Morita, "Specific Features of a Deductive Object-Oriented Database Language $\mathcal{QUIXOTE}$," *Proc. ACM SIGMOD Workshop on Combining Declarative and Object-Oriented Databases (SIGMOD'93 WC-DOOD)*, pp.89-99, Washington DC, USA, May 29, 1993.

Yokota et al 01) Kazumasa Yokota, Takeo Kunishima, and Bojiang Liu, "Semantic Extensions of XML for Advanced Applications," *Australian Computer Science Communications, Volume 23, Number 6 (Proc. Workshop on Information Technology for Vietual Enterprises (ITVE 2001))*, pp.49-57, Jan., 2001.

11

XML Databases

Masatoshi Yoshikawa

Nara Institute of Science and Technology

ABSTRACT

With rapid development of standards and supporting software tools, XML (eXtensible Markup Language) is becoming pervasive as the next generation Web language. Since XML is widely attracting attention as a language for describing data as well as structured documents, XML's role in Web databases is crucial. In this paper, we describe major research issues on the interrelation between XML and database technologies. First, we present the data model of XML1.0. Then, we provide an overview of query languages for XML. Next, we describe various approaches to the storage and retrieval of XML data. In particular, we focus on alternative methods of mapping XML data into relational database schemas. Fast search of XML data requires special physical data organizations. Finally, we present indices for XML data, and show how each index support fast retrieval for representative query operations.

1 Introduction

XML(eXtensible Markup Language) is becoming widely used as a standard meta language to represent structured documents and data on the Web. One of the major reasons for the wide acceptance of XML is its independence from platforms, operating systems, network protocols and applications. The wide dissemination of the Web and the advent of universal data format such as XML have a crucial impact on the research of databases. The research issues raised by the interrelationship between XML and database can be summarized in the following two categories:

1. **Database management systems as repositories for XML documents:** Since XML has now become a standard language, XML documents will be produced in wide area of applications. Efficient and flexible management of large volume of XML documents requires the functionalities of DBMSs such as fast retrieval and update, integrity constraints enforcement, concurrency control, access right control, and version management. An important issue is how conventional DBMSs or text indices can be used, or can be adapted to cope with large volume of XML data. Unlike many other types of data stored in traditional databases, XML documents has logical structures. The development of techniques to exploit such logical structures as granules for the use of

functionalities of DBMSs is an important research issue. If XML documents are stored in relational or object-oriented databases, the development of algorithms for translating queries in XML query languages into SQL or OQL is also a mandatory technical issue.

2. **XML as a common data model on the Web:**

   A huge amount of information in various format is available on the Web. While multimedia data is a powerful means to convey information intuitively, a significant part of intellectual information on the Web is presented in forms of text. Some of those data are stored in relational or object-oriented databases, while some others are stored in HTML or plain text files.

   If there is a common data model which is capable of representing such wide variety of text based data, we could transform existing data on the Web into the common data model. The transformed data does not need to be materialized, but could be a view of existing data. With the existence of a common data model, data on the Web becomes uniformly accessible, and thus interchange and distribution of information is greatly facilitated. XML is expected to become such a common data model on the Web. Here, a natural and important question is: "Does XML have enough power to express wide variety of existing information resources on the Web ?" The XML1.0 data structure has enough expressive power in that it can simulate relational data model and a significant part of object-oriented data model. Also, HTML is a language whose data structure could be defined by XML DTD. However, the current XML1.0 lacks of data types such as integer, real and date. A set of extensive data types will be supported in schema languages under development such as XML Schema[20, 21] and RELAX[13].

   By transforming existing data into XML, the Web can be regarded as a huge, highly-distributed XML database. The notion of "database" here is not same as those of traditional DBMSs. Databases in this broader sense means a collection of physical or virtual XML documents.

2 XML1.0

XML1.0[23] is a core standard in the family of XML related standards. To put it simply, a DTD of XML1.0 is a context-free grammar, and an XML document conforming to the DTD is a parse tree. XML Information Set[27] describes an abstract information set available in an XML document. Also, DOM (Document Object Model)[22], an API (Application Programming Interface) for XML documents, and XPath (XML Path Language)[26] respectively defines their own data models. In this paper, we will follow the XPath data model. Wadler gave a formal description of the XPath data model[17].

For example, the XML document in Figure 1 can be represented as a tree in Figure 2. In this figure, triangle nodes, circle nodes and rectangular nodes represents

```
<readinglist courseid="CS405">
        <book isbn="0-07-232206-3">
                <title>Database Management Systems</title>
                <authors>
                        <author>
                                <first>Raghu</first>
                                <last>Ramakrishnan</last>
                                <affiliation>University of Wisconsin
                                          - Madison</affiliation
                        </author>
                        <author>
                                <first>Johannes</first>
                                <last>Gehrke</last>
                        </author>
                </authors>
                <summary>This book covers the fundamentals of
                        modern database management systems, ...
                </summary>
        </book>
        <paper>
                <title>The Database Language GEM</title>
                <authors>
                        <author>
                                <first>Carlo</first>
                                <last>Zaniolo</last>
                        </author>
                </authors>
                <booktitle>SIGMOD'83</booktitle>
                <abstract>GEM is a general-purpose query and
                        update language for ...
                </abstract>
        </paper>
</readinglist>
```

Figure 1: An XML document.

element nodes, attribute nodes and document nodes, respectively.

3 Query Languages

Proposal of query languages has been made for the last couple of years. After intensive discussion at Query Language Workshop in December 1998[24)], W3C has organized the XML Query Working Group. The goal of the Working Group is to produce a data model for XML documents, a set of query operators, and a query language based on these query operators[28)]. The data model will be based on the XML Infoset[27)], and will support Namespace[25)]. Requirements for the data model include the support for a collection of documents and references.

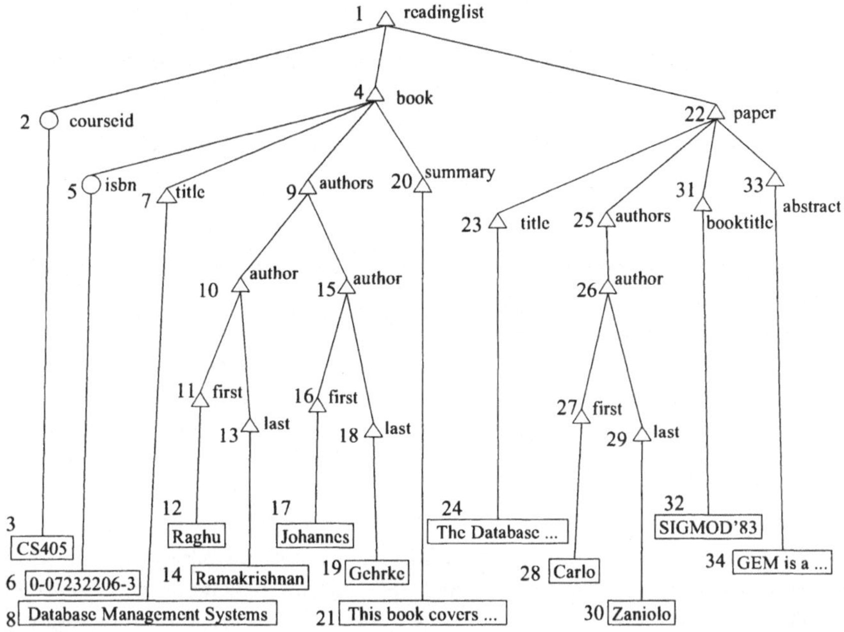

Figure 2: A tree representation of an XML document.

3.1 Requirements for Query Languages

Although the W3C XML Query Working Group is still in the mid of standardization process, general consensus about the requirements for query languages are emerging[2)12)28)]:

- The query language should be declarative so that queries are optimized by systems.
- The query language may have more than one syntax bindings. Possible syntax include XML, and ones which can be embedded in URL. Human-readability is also an important factor. Another requirement is the convenience for programs to generate and manipulate queries.
- The query language should support for intra- and inter-document references. XLink[18)] and XPointer[19)] can be used to specify references.
- Queries should be possible for documents without a schema. Furthermore, query should be possible for accessing the schema of a document, if it is available.

- The query language should be closed with respect to the data model. Namely, results of queries should be an XML document or a set of XML documents.
- The query language should support for the transformation of the structure of documents, and full text retrieval.
- The query language should be able to create new documents.

3.2 XML-QL

Until now, several query languages for XML documents have been proposed. Bonifati and Ceri have made comparative study of five languages[1)].

In the rest of this section, we will see how queries are represented in XML-QL[5, 6)]. The basic constructs of the XML-QL syntax are `where` clause and `construct` clause. `Where` clause specifies filtering and variable binding, whereas `construct` clause specifies the structure of query results. For example, the query in Figure 3(a) returns authors and title of papers whose `booktitle` element value is "SIGMOD'83" for the XML document in Figure 1. Although the condition in `where` clause is also expressible in XPath, XPath do not have the ability of the `construct` clause of this query which extracts elements and reorder them. XPath's "//" operator can be expressed by specifying wild card "$" and its Kleene star "*" as element name. For example, Figure 3(b) shows a query returns `authors` elements which appear in documents. The relational join operator is expressed by placing copies of a variable in positions of joined values. For example, the query in Figure 3(c) lists up researchers who wrote one or more papers and one or more books. In databases supporting object identity, there is an idea of the use of Skolem functor for creating new objects[11)9)]. Following this idea, transformation of XML documents can be expressed. For example, the query in Figure 3(d) creates a `person` element for each pair of bounded values of `$F` and `$L`. In the query, `personID` plays the role of a Skolem functor.

4 Storing XML Documents in Databases

The development of technologies for storing and retrieving XML documents is an important research issue. Until now, many proposals have been made. These proposals can be roughly classified into the following three methods:

1. use of relational databases
2. use of object-oriented databases
3. development of dedicated XML repository

For each of the three methods, commertial XML database systems are available. We believe the use of relational databases is more practical than other two methods for the following reasons:

```
where <paper>
        <booktitle>
          SIGMOD'83
        </booktitle>
        <title>$T</title>
        <authors>$A</author>
      </>
construct <sigmod83paper>
        <authors>$A</>
        <title>$T</>
      </>
```

(a)

```
where <$*>
        <authors>$A</author>
      </>
construct <authors>$A</>
```

(b)

```
where <$*>
        <book><authors><author>
          <first>$F</>
          <last>$L</>
        </></></>
        <paper><authors><author>
          <first>$F</>
          <last>$L</>
        </></></>
      </>
construct <researcher>
          <firstname>$F</>
          <lastname>$L</>
        </>
```

(c)

```
where <$*>
        <authors><author>
          <first>$F</>
          <last>$L</>
        </></>
        <title>$T</>
      </>
construct
   <person ID=personID($F,$L)>
       <firstname>$F</>
       <lastname>$L</>
       <title>$T</>
   </>
```

(d)

Figure 3: XML-QL queries.

- Query optimization and query processing in relational databases is a matured technology which is the result of accumulation of improvements and modifications during the past quarter century. Methods using relational databases can enjoy such valuable technological assets.
- Relational databases are already used in many organizations. Storing XML documents in relational databases facilitates integrated usages of XML documents and other data.

In this section, we present several approaches to storing XML documents in relational databases. These approaches can be roughly classified into methods for designing relational schema based on DTDs, and methods for mapping XML documents into relational tables without the knowledge of DTDs. In both approaches, queries on XML documents are translated into SQL queries on the underlying relational databases.

4.1 A Method for Designing Relational Schema based on DTDs

Shanmugasundaram et. al. has proposed a method to design relational database schemas based on the analysis of DTDs of XML documents[15)]. Let us assume that the XML documents in Figure 1 is conformant to the DTD in Figure 4. In their

```
<!ELEMENT readinglist (book*, paper*)>
<!ATTLIST readinglist courseid ID REQUIRED>
<!ELEMENT book (title, authors, summary?)>
<!ATTLIST book isbn ID REQUIRED>
<!ELEMENT paper (title, authors, booktitle, abstract?)>
<!ELEMENT title (#PCDATA)>
<!ELEMENT authors (author+)>
<!ELEMENT author (first, last)>
<!ELEMENT first (#PCDATA)>
<!ELEMENT last (#PCDATA)>
<!ELEMENT summary (#PCDATA)>
<!ELEMENT booktitle (#PCDATA)>
<!ELEMENT abstract (#PCDATA)>
```

Figure 4: DTD for the XML document in Figure 1.

method, DTDs simplification is performed as a initial step. For each child element e of elements, information on the number of occurrences (i.e. once or more than once) of e and information whether e is mandatory or not is retained. However, information regarding the occurrence order of distinct elements is discarded. A graph representing DTDs after simplification is called a DTD graph. For example, the DTD graph of the DTD in Figure 4 is shown in Figure 5.

Two major methods, called *Shared* and *Hybrid*, are proposed for translation of DTD graphs into relatinal database schemas. In the *Shared* method, for each element with indegree greater than or equal to 2 in DTD graphs (i.e. element shared by more than one elements), a relation schema is created. Elements with indegree 1 are translated into an attribute of the relation corresponding to an ancestor element. Also, a relational schema is created for each elements with indegree 0. Here, for each element represented by a destination node of an edge labeled with "*", a separate relational schema is created. This is because set values by themselves cannot be stored in relational databases. Furthermore, elements which are reachable along directed paths in DTD graphs from an element having its own relational schema, say R, are "inlined" into R (i.e. they are defined as attriutes of R.), provided that the directed paths do not include an edge labeled with "*". For example, the *Shared* translates the DTD graph in Figure 4 into the relational database schema in Figure 6(a). In the *Hybrid* method, elements with indegree greater than or equal to 2 are also inlined, if they are reachable from other elements without passing an edge with "*". The *Hybrid* method yields the relational database schema in Figure 6(b).

The information on the occurrence order of elements, which was discarded at itital step, is recovered by adding the occurrence position of elements in relational database schemas.

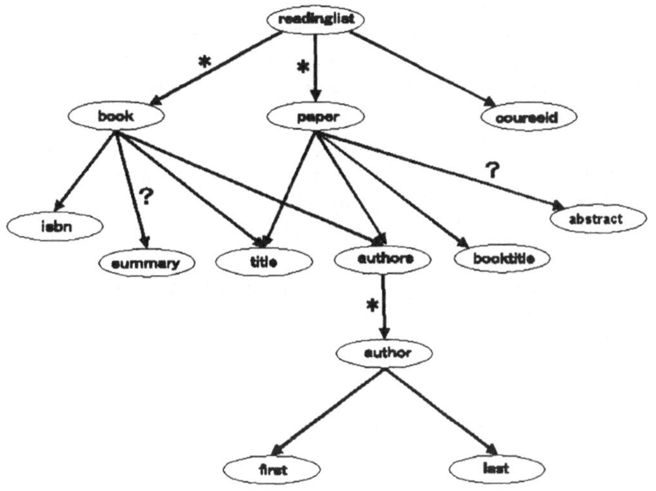

Figure 5: A DTD graph.

4.2 Methods for Mapping XML Documents into Relational Databases without the Knowledge of DTDs

Florescu and Kossmann reported the result of performance evaluation of several alternative ways to store XML documents into relational databases[7)]. The alternative ways they studied do not use the information of DTDs. XML documents are modeled as a tree as shown in Figure 2. For simplicity, the distinction of elements and XML attributes is blurred. Florescu and Kossmann identified two axes for classifying mappings of tree data into relational schemas: one is mapping edges and the other is mapping values. They listed up the following three approaches to mapping edges:

(E_e) Edge Approach

A relation is created for storing information of edges in a tree. Each edge is stored as a tuple in the relation.

(E_b) Binary Approach

Edges are clusterd according to element names (or XML attribute names) of their destination nodes. A relation is created for each cluster.

(E_u) Universal Table

A relation is created which have attributes for every elements and XML attributes in the target XML document.

```
readinglist(readinglistID: integer,
          readinglist.courseid: string)
book(bookID: ingeger, book.parentID: integer,
     book.isbn: string, book.summary, string)
paper(paperID: integer, paper.parentID: integer,
      paper.booktitle: string, paper.abstract: string)
title(titleID: integer, title.parentID: integer,
     title.parentCODE: integer, title: string)
authors(authorsID: integer, authors.parentID: integer,
        authors.parentCODE: integer)
author(authorID: integer, author.parentID: integer,
       first: string, last: string)
```

(a)

```
readinglist(readinglistID: integer,
          readinglist.courseid: string)
book(bookID: ingeger, book.parentID: integer,
     book.isbn: string, book.summary: string,
     book.title: string, book.authors.authorsID: integer)
paper(paperID: integer, paper.parentID: integer,
      paper.booktitle: string, paper.abstract: string,
      paper.title:string, paper.authors.authorsID: integer)
author(authorID: integer, author.parentID: integer,
       author.parentCODE: integer, first: string,
       last: string)
```

(b)

Figure 6: Relational database schemas.

As for mapping values, the following two approaches are listed:

(V_s) Separate Value Tables

Relations are created for each data types.

(V_i) Inlining

Values are inlined into relations created by the "mapping edges" approach.

Theoretically, any combination of the three approaches to mapping edges and two approaches to mapping values is possible. For example, Figure 7 shows the relational schema resulting by combining ($\mathbf{E}_e$) and ($\mathbf{V}_s$). Also, Figure 8 gives the combination of ($\mathbf{E}_b$) and ($\mathbf{V}_i$). As a result of performance evaluation, Florescu and Kossmann concluded that the combination of ($\mathbf{E}_b$) and ($\mathbf{V}_i$) is the best for retrieval.

Other approaches are also possible. We have proposed the following approach[16, 30] which is on the "mapping edge" axis in Florescu and Kossmann's classification.

Edge

source	ordinal	name	target
...	...	...	...
4	2	title	v_8
...	...	...	...
9	1	author	10
9	2	author	15
...	...	...	...
22	1	title	v_{24}
...	...	...	...
25	1	author	26
...	...	...	...

Value

vid	value
...	...
v_8	Database Management Systems
...	...
v_{24}	The Database Language GEM
...	...

Figure 7: The edge approach with separate value table.

title

source	ordinal	val_{int}	val_{string}	target
4	2	null	Database Management Systems	null
22	1	null	The Datase Language GEM	null

author

source	ordinal	val_{int}	val_{string}	target
9	1	null	null	10
9	2	null	null	15
25	1	null	null	26

⋮

Figure 8: The binary approach with inlined values.

(E_p) Path Approach

Trees representing an XML document are decomposed into paths from the root to other nodes. Then, those paths are stored in one relation as a string. Also, three more relations are created for storing element nodes, XML attribute nodes, and text nodes, respectively.

The path approach is classified into the inlining on the "mapping value" axis. For example, the XML document in Figure 1 is mapped into the relations in Figure 9 in the path approach.

4.3 A Comparison of Alternative Methods

Table 1 gives a comparison of alternative methods for storing XML documents in relational databases.

Element

docID	*pathID*	*index*	*reindex*	*pos*
1	p1	0	-1	0.1, 149.3
1	p3	0	-1	0.2, 149.2
1	p5	0	-1	0.3, 3.1
1	p6	0	-2	3.4, 4.1
...	...	...	...	... ...
1	p6	1	-1	10.4, 11.1
...	...	...	...	... ...
1	p18	0	-1	98.2, 149.1

Attribute

docID	*pathID*	*attvalue*	*pos*
1	p2	CS405	0.1, 0.1
1	p4	0-07232206-3	0.2, 0.2

Text

docID	*pathID*	*value*	*pos*
1	p5	Database Management System	1, 3
1	p6	Raghu	4, 4
...	...	...	...
1	p6	Johannes	11, 11
...	...	...	...
1	p18	GEM is a ...	99, 149

Path

pathexp	*pathID*
/readinglist	p1
/readinglist/@couseid	p2
/readinglist/book	p3
/readinglist/book/@isbn	p4
/readinglist/book/title	p5
/readinglist/book/authors/author/first	p6
...	...
/readinglist/paper/abstract	p18

Figure 9: The path approach.

In Shanmugasundaram's method, the binary approach and the universal table approach, database schema is dependent on the logical structure of XML documents. Hence, in these methods, update of XML documents or insertion of new XML documents may cause re-design of database schema.

Path expressions often appear in XML queries. In the path approach, paths are stored as character string, hence, processing of path expressions can be made by string comparison in SQL. However, other methods requires join operations in proportion to the length of path expressions.

5 Indices for XML Documents

Indices for XML documents should have a data structure which is suitable for fast access to document fragments based on document logical structures and/or character strings.

According to Sacks-Davis[14)], indexing of structured documents can be classified into the following two approaches:

- position-based indexing, and

Table 1: A comparison of methods of storing XML documents in relational databases.

DTD	Method		dependence of database schema on XML document structure	Join operations proportional to path length
	Mapping			
Use	Shanmugasundaram's method		$\surd$	$\surd$
No Use	Edge	(E_e) Edge		$\surd$
		(E_b) Binary	$\surd$	$\surd$
		(E_u) Universal Table	$\surd$	$\surd$
		(E_p) Path		
	Value	(V_s) Separate Tables	N/A	$\surd$
		(V_i) Inlining	N/A	

- path-based indexing.

These two approaches are not mutually exclusive. In fact, there are hybrid indices which own the both characteristics.

In position-based indices, the position of the occurrences of text objects (such as tags, attributes, characters, and etc.) are represented by the offset from the first character of documents. Occurrences of elements are represented by a pair of their start and end positions[4, 3], which is called a *region*. Another method of representing regions, called Relative Region Coordinate, is proposed to efficiently handle updates[10]. In processing queries on XML documents, it is often necessary to retrieve occurrence positions of a given element containing a given word. Such operations are easily processed by using invert indices for elements and inverted indices for words.

In path-based indices, occurrence positions of elements and attributes in a document are represented by path expressions. Usually, paths from the root to element (or attribute) nodes in concern are used. In many queries, however, conditions on paths are specified ambiguously. We have proposed a new index which is suitable for efficient processing of ambiguous path expressions[29]. In our index, for each element node n which has mixed content, a string concatenating

1. start tags from the root to n,
2. the content of n, and
3. end tags from n to the root

is created. We call such concatenated string an ENRP (Element with Normal and Reverse Path). For each attribute node, a character string called ANRP (Attribute with Normal and Reverse Path) is created analogously. The new index is basically a suffix array[8] of ENPRs and ANPRs. Let us consider the document tree in Figure 2.

```
ENRP for the element node 8:
     <readinglist><book><title>Database Management Systems</title>
     </book></readinglist>
ENRP for the element node 24:
     <readinglist><paper><title>The Database Language GEM</title>
     </paper></readinglist>

ANRP for the attribute node 5:
     <readinglist><book>@isbn 0-07-232206-3 @isbn</book></readinglist>
```

Figure 10: ENRPs and ANRPs

Figure 10 gives the ENPRs for the element nodes 8 and 24, and the ANPR for the attribute node 5. The index enjoys the advantages of suffix arrays. Furthermore, the index facilitates fast access to documents based on logical structures. The sequence of end tags in ENRPs and ANRPs are effective in efficiently processing ambiguous path expressions, such as '//title' in XPath.

Acknowledgments

The author thanks Dr. Toshiyuki Amagasa, Mr. Dao Dinh Kha, Mr. Takeyuki Shimura and Mr. Yohei Yamamoto for their development of the XML database and index systems mentioned in this paper. The author is grateful to Professor Shunsuke Uemura at Nara Institute of Science and Technology for his continuous encouragement.

Bibliography

1) Angela Bonifati and Stefano Ceri. Comparative analysis of five xml query languages. *SIGMOD Record*, 29(1):68–79, 2000.

2) Paul Cotton and Ashok Malhotra. Summary of requirements gleaned from workshop position papers. QL'98 - The Query Languages Workshop, November 1998. http://www.w3.org/TandS/QL/QL98/pp/queryreq.html.

3) Tuong Dao. An indexing model for structured documents to support queries on content, structure, and attributes. In *Proc. of the IEEE International Forum on Research and Technology Advances in Digital Libraries (ADL'98)*, pages 88–97, April 1998.

4) Tuong Dao, Ron Sacks-Davis, and James A. Thom. An indexing scheme for structured documents and its implementation. In *Proc. of the 5th International Conference on Database Systems for Advanced Applications (DASFAA'97)*, April 1997.

5) Alin Deutsch, Mary Fernandez, Daniela Florescu, Alon Levy, and Dan Suciu. XML-QL: A Query Language for XML. http://www.w3.org/TR/NOTE-xml-ql/, August 1998.

6) Alin Deutsch, Mary Fernandez, Daniela Florescu, Alon Levy, and Dan Suciu. A Query Language for XML. *WWW8 / Computer Networks*, 31(11-16,17):1155–1169, May 1999.

7) Daniela Florescu and Donald Kossmann. Storing and querying xml data using an rdmbs. *IEEE Data Engineering Bulletin*, 22(3):27–34, September 1999.

8) Dan Gusfield. *Algorithms on Strings, Trees, and Sequences*. Cambridge University Press, 1997.

9) Richard Hull and Masatoshi Yoshikawa. Ilog: Declarative creation and munipulation of object identifiers. In *Proc. of the 16th International Conference on Very Large Data Bases (VLDB)*, pages 455–468, Brisbane, Aug. 1990.

10) Dao Dinh Kha, Masatoshi Yoshikawa, and Shunsuke Uemura. An xml indexing structure with relative region coordinate. In *Proc. of IEEE 17th International Conference on Data Engineering*, pages 313–320, April 2001.

11) David Maier. A logic for objects. In *Workshop on Foundations of Deductive Databases and Logic Programming*, pages 6–26, 1986.

12) David Maier. Database Desiderata for an XML Query Language. In *Position papers for W3C Query Language Workshop*. December 1998. http://www.w3.org/TandS/QL/QL98/pp/maier.html.

13) Makoto Murata. Relax (regular language description for xml). http://www.xml.gr.jp/relax/.

14) Ron Sacks-Davis, Tuong Dao, James A. Thom, and Justin Zobel. Indexing documents for queries on structure, content and attributes. In *International Symposium on Digital Media Information Base (DMIB'97)*, pages 236–245. World Scientific, 1998.

15) Jayavel Shanmugasundaram, Kristin Tufte, Gang He, Chun Zhang, David J. DeWitt, and Jeffrey F. Naughton. Relational Databases for Querying XML Documents: Limitations and Opportunities. In Malcolm P. Atkinson, Maria E. Orlowska, Patrick Valduriez, Stanley B. Zdonik, and Michael L. Brodie, editors, *VLDB'99, Proceedings of 25th International Conference on Very Large Data Bases, September 7-10, 1999, Edinburgh, Scotland, UK*, pages 302–314. Morgan Kaufmann, 1999.

16) Takeyuki Shimura, Masatoshi Yoshikawa, and Shunsuke Uemura. Storage and Retrieval of XML Documents using Object-Relational Databases. In *Proc. of the 10th International Conference on Database and Expert Systems Applications (DEXA'99)*, volume 1677 of *Lecture Notes in Computer Science*, pages 206–217. Springer-Verlag, August-September 1999.

17) Philip Wadler. A formal semantics of patterns in xslt. In *Proc. of Markup Technologies '99*, pages 1–11. GCA, December 1999.

18) World Wide Web Consortium. XML Linking Language (XLink) Version 1.0. http://www.w3.org/TR/xlink/.

19) World Wide Web Consortium. XML Pointer Language (XPointer) version 1.0. http://www.w3.org/TR/xptr.

20) World Wide Web Consortium. XML Schema Part 1: Structures. http://www.w3.org/TR/xmlschema-1. W3C Working Draft.

21) World Wide Web Consortium. XML Schema Part 2: Datatypes. http://www.w3.org/TR/xmlschema-2. W3C Working Draft.

22) World Wide Web Consortium. Document Object Model(DOM) level 1 specification version 1.0. http://www.w3.org/TR/1998/REC-DOM-Level-1-19981001, October 1998. W3C Recommendation 1-October-1998.

23) World Wide Web Consortium. Extensible Markup Language (XML) 1.0. http://www.w3.org/TR/1998/REC-xml-19980210, February 1998. W3C Recommendation 10-February-1998.

24) World Wide Web Consortium. QL'98 - The Query Languages Workshop. http://www.w3.org/TandS/QL/QL98/, December 1998.

25) World Wide Web Consortium. Namespaces in XML. http://www.w3.org/TR/1999/REC-xml-names-19990114/, January 1999. W3C Recommendation 14-January-1999.

26) World Wide Web Consortium. XML Path Language (XPath) version 1.0. http://www.w3.org/TR/xpath, November 1999. W3C Recommendation 16 November 1999.

27) World Wide Web Consortium. XML Information Set. http://www.w3.org/TR/2000/WD-xml-infoset-20000726, July 2000. W3C Working Draft 26 July 2000.

28) World Wide Web Consortium. XML Query Requirements. http://www.w3.org/TR/xmlquery-req, February 2001. W3C Working Draft 15 February 2001.

29) Yohei Yamamoto, Masatoshi Yoshikawa, and Shunsuke Uemura. On Indices for XML Documents with Namespaces. In *Markup Technologies'99*, pages 235–243, Philadelphia, U.S.A., December 1999. GCA.

30) Masatoshi Yoshikawa, Toshiyuki Amagasa, Takeyuki Shimura, and Shunsuke Uemura. Xrel: A path-based approach to storage and retrieval of xml documents using relational databases. *ACM Transactions on Internet Technology*, 1(1):110-141, June 2001.

12

Construction of Web Structures from Heterogeneous Information Sources

Hiroyuki Kitagawa
Atsuyuki Morishima

Institute of Information Sciences and Electronics
University of Tsukuba

ABSTRACT

With the broad acceptance of the World Wide Web, the Web has been widely used for publishing and disseminating information originally stored in various information sources. They include databases, document repositories, and Web servers. In constructing Web structures on top of the heterogeneous information sources, data acquisition is an essential issue. Also, design of data layout on Web pages is another important issue. A lot of work has been reported on the problem so far, and some tools and systems are used in practical applications. In this article, we survey the current approaches to this problem in the first part, and show our approach in the second part. The current approaches to the problem can be classified into the program development approach, the template-based approach, and the mediation approach. In the first approach, application programs to acquire data and to generate Web pages are developed from scratch, and they are invoked through CGI, Servlet, and/or Web server API. In the template-based approach, some kinds of layout templates with scripts and queries are used to specify the data acquisition and data layout simultaneously. In the mediation approach, a common mediation data model is introduced to specify layout-independent logical data acquisition or data integration. In this article, we mainly explore the mediation approach, since this approach is most promising in the context of large-scale Web structure construction. We give an overview of the current technology related to this approach from a number of important design viewpoints. The second part of this article explains some details of our approach. It takes the mediation approach. The common data model is based on nested relations and ADTs, and interactive visual authoring facilities are provided for data layout specification.

1 Introduction

With the broad acceptance of the World Wide Web, the Web has been widely used for publishing and disseminating information originally stored in various information sources. They include databases, document repositories, and Web servers. Needless to say, in constructing Web structures on top of the heterogeneous information sources, data acquisition from the sources is an essential issue. It usually involves querying databases, accessing data files, restructuring the obtained data, and so on. Moreover, the Web structure construction poses us another important problem. It is the Web page layout design. This aspect is one of the factors that makes the construction of Web structures different from the conventional data integration on heterogeneous information sources.

In the first part of this article, we overview the current approaches to the Web structure construction problem. We classify them into the program development approach, the template-based approach, and the mediation approach. Among the three approaches, the mediation approach is most promising for the construction of large-scale Web structures. Therefore, we focus on the mediation approach, and discuss a number of important design issues related to this approach. The second part of this article explains some details of our approach, which follows the mediation approach.

The first part of this article consists of Sections 2 and 3. In Section 2, we give a reference model for the Web structure construction problem, and classify the current approaches into the three categories. In Section 3, we explore the mediation approach, and discuss a number of important design issues and current approaches. Section 4 corresponds to the second part, and introduces some details of our approach to this problem. Section 5 is the conclusion.

2 Reference Model and Three Major Approaches

2.1 Reference Model of the Web Structure Construction Task

In this subsection, we give a reference model of the Web structure construction task. In general, the task can be divided into two processes. The first one is *Data Acquisition* and the second one is *Data Layout* (Figure 1). The data acquisition is the process to extract data required for constructing target Web structures from underlying information sources. The information sources may be conventional databases, legacy data files, document repositories, multimedia data files, and Web pages. Data acquisition includes data selection, grouping, aggregation, and reformatting. Data layout is required to map the extracted data to the Web structure. It is often the case that in this process we add some decorative objects such as title banners to or remove unnecessary data objects from the extracted data.

2.2 Three Major Approaches

There are a number of approaches to the Web structure construction task. They can be classified into three approaches: (1) the program development approach, (2) the template-based approach, and (3) the mediation approach.

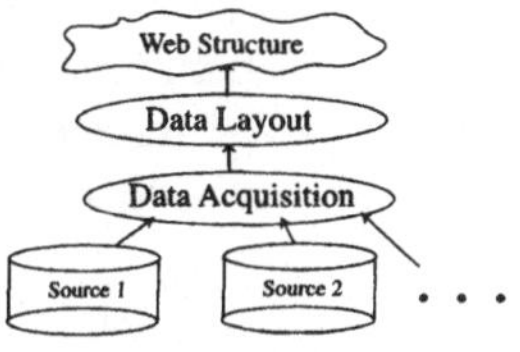

Figure 1: Reference model of the Web structure construction task

(1) Program Development Approach

This approach is the most primitive one. Many projects to develop practical systems have taken this approach so far. In this approach, both the data acquisition and the data layout are performed in hand-coded application programs (Figure 2(a)). Those programs are written in general-purpose programming and scripting languages such as perl and java. They usually include access requests to the information sources and database queries. They have to generate HTML documents as their output. Those tailor-made application programs are invoked through frameworks such as CGI and Servlet[44)], and Web Server APIs such as ISAPI[33)] and NSAPI[38)]. When invoked through CGI, programs are associated with URLs.

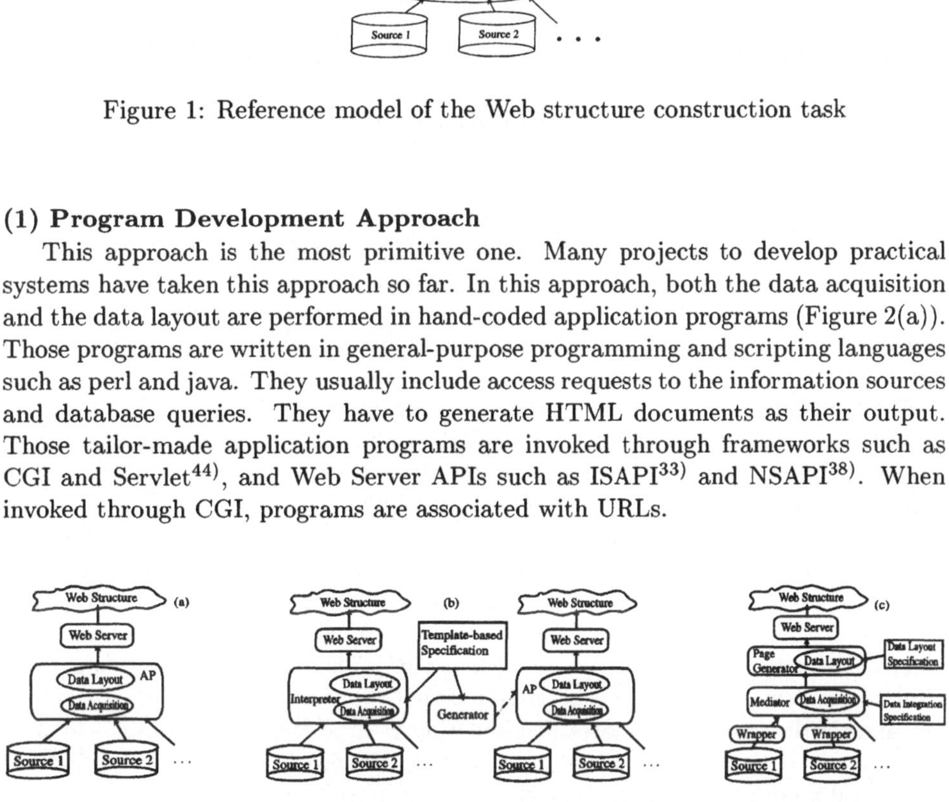

Figure 2: Three major approaches

(2) Template-based Approach

Since the Web structure construction problem has been recognized as an important issue for application development, more sophisticated approaches have emerged. They include the template-based approach (Figure 2(b)). In this approach, templates are used to give data layout specification. Typically, the templates are based on HTML. Data acquisition specification is given as queries and scripts embedded into those templates. The approach is similar to the program development approach in that the data acquisition and the data layout are specified together. However, it has a number of advantages over the program development approach.

In this approach, users need not write programs from scratch. In particular, construction of templates is far easier compared with writing program codes to generate HTML documents. Another advantage is that construction of templates can be regarded as a natural extension of authoring ordinary static Web pages. Well-known examples of the template-based approach include Microsoft's Active Server Pages[33] and Allaire Inc.'s ColdFusion[5]. DB2WWW connection[39] also follows this approach. Embperl[46] is often used to write scripts embedded in HTML templates.

(3) Mediation Approach

Another approach is the mediation approach. In the mediation approach, a common mediation data model is introduced to facilitate logical data acquisition process. Usually, integration of data from the underlying information sources is attained based on the common data model. Thus, the data acquisition process can be regarded as a data integration process in this approach. In this approach, the data acquisition and data layout are processed by different components (Figure 2(c)). Typically, the data acquisition is performed by the mediator[47] and wrappers. Wrappers provide the mediator with views of the underlying information sources in the common data model. The mediator analyzes the data integration specification, decomposes it into local data acquisition requests, and sends them to the wrappers. The wrapper receives the intermediate result, translates it into the common data model, and sends it back to the mediator. The mediator collects data from the wrappers and produces the data acquisition result. The page generator receives the data layout specification, and maps the data acquisition result into the Web structures.

An advantage of the mediation approach is that it is suitable for construction of large-scale Web structures. One of the reasons is that the integrated data resulted from the data acquisition process can be used to construct a collection of Web pages which links to each other. In contrast, a template used in the template-based approach is usually associated with a particular URL. Therefore, users need not write similar queries many times. Another reason is that the separation of the data acquisition and the data layout facilitates manipulation of large-scale data. Systems following this approach include Strudel[18], Tiramisu[6], Araneus[8], and InfoWeaver[26, 35].

Several studies have shown that the mediation approach is most promising for large-scale Web structure construction based on heterogeneous information sources[8, 19]. For this reason, we further discuss design issues related to the mediation approach in Section 3. In Section 4, we explain the Web structure construction in InfoWeaver as an example of the mediation approach. InfoWeaver provides visual support for both the data acquisition and data layout specification.

3 Design Issues Related to the Mediation Approach

3.1 Common Data Model for Data Integration

Common data models are key components for the data acquisition in the mediation approach. Here, we classify them into three categories: semistructured, well-structured, and hybrid data models. We explain their differences using the example data shown in Figure 3. In this example, we assume that a relational database contains phone number information, while address information is stored in XML documents. Note that address items in the XML documents have slightly irregular structure. The zip codes are placed at different positions.

(1) Semistructured data models

Semistructured data models are commonly used in the context of data integration. A lot of systems follow this approach to attack the Web structure construction on heterogeneous information sources[6, 7, 9, 19, 21]. In this approach, instance-based schema-less graph data models [1, 2, 3, 11, 40] are often used. Similar models are often used to model XML documents[49]. An advantage of this approach is that it is very flexible and can represent various data structures in a single uniform framework (Figure 4). TSIMMIS[40] uses the OEM data model to integrate information sources. Strudel[18] is a Web site management system, and uses a similar data model. WebOQL[7] is a Web query language based on a variation of graph-based semistructured data models. MIX[9] is a mediator system which directly uses XML as the common data model. MIX uses the query language called XMAS.

SSN	name	tel
12	Diane	45-6789
34	Vincent	23-2345

```
<addr-list>
<item><name>Diane</name>
  <addr><street>Hobart</street><state>NJ</state><zip>67</zip></addr>
</item>
<item><name>Vincent</name>
  <zip>45</zip>
  <addr><street>Main</street><state>NY</state></addr>
</item>
</addr-list>
```

Figure 3: Example data

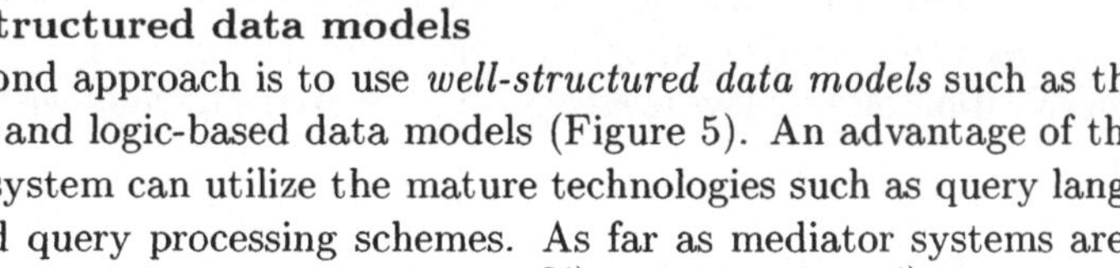

Figure 4: Graph-based semistrucuted data model view

(2) Well-structured data models

The second approach is to use *well-structured data models* such as the relational data model and logic-based data models (Figure 5). An advantage of this approach is that the system can utilize the mature technologies such as query languages, data storage, and query processing schemes. As far as mediator systems are concerned, Disco [45] uses relations, and Infomaster [24] and HERMES [4] use logic-based representations. Araneus[8] can extract information from the Web pages and produce new Web structures based on relational structures. WebLog[28] is a logic-based language which enables us to query and restructure Web pages, exploiting various extensible built-in predicates. A drawback of this approach is that the mapping of heterogeneous information into the well-structured data model is not generally easy.

SSN	name	tel
12	Diane	45-6789
34	Vincent	23-2345

name	street	state	zip
Diane	Hobart	NJ	67
Vincent	Main	NY	45

Figure 5: Relation-based well-strucuted data model view

(3) Hybrid data models

The third approach is the *hybrid* approach. In this approach, data modeling frameworks fit for different information sources are combined. For example, ADTs are often incorporated into relational frameworks to accept semistructured data such as XML documents (Figure 6). An advantage is that development of wrappers is relatively easy. MIROWeb[10] introduces the semistructured data types to accept XML documents and XML-QL[49] queries. Another possible advantage is that we can exploit various functions provided by ADT methods. InfoWeaver[26, 35, 36] features dynamic conversion between relational data and the document type, and allows users to apply ADT methods to the document data. The hybrid model can be constructed based on the object model rather than the relational model. In this case, ADTs correspond to object types in the object model.

SSN	name	tel
12	Diane	45-6789
34	Vincent	23-2345

addr-list
"<addr-list><item>...</item> <item>...</item></addr-list>"

Figure 6: Relation-and-ADT-based hybrid data model view

When we integrate data in different information sources, the system has to resolve their structural heterogeneity. Facilities for this process heavily depend on characteristics of the common data model. Well-structured data models usually require that the heterogeneity be completely resolved when data is mapped into the common data model. For example, WebLog's users have to decide how to map Web information to the simple data structure it can deal with. Araneous provides a language for mapping hypertext structures into relations. On the other hand, in systems based on semistructured and hybrid data models, this resolution is "postponed" until the user describes the integration specification. In this case, languages for the integration specification often allow *regular path expressions* to cope with structural heterogeneity.

3.2 Data Layout Specification

As mentioned in Section 1, data layout is also essential as well as the data acquisition. A well-known example of layout specification languages is the HTML-based template language for Strudel[19]. Also, existing Web page authoring tools can be used for the data layout specification[6]. Authoring tools such as FrontPage[33] and

DreamWeaver[32] allow users to drag and drop visual objects into blank windows to realize WYSIWYG and interactive layout specification environment.

3.3 Visual Environments

Since visual aspects are important in the Web page design, it is natural to provide visual environments to support this. As explained in Subsection 3.2, there are many Web page authoring tools that can be used for the layout specification.

Some visual facilities can support the data integration specification. QBE[50] and its variations are widely used for visual queries to the relational data. Semistructured data can be manipulated by visual facilities, too. DataGuide[25] is an abstract specification of semistructured data, and can be used as a QBE-like query language for semistructured data. XML-GL[12] is a graph-based query language for XML documents. BBQ is a visual interface for XMAS query specification in MIX. An example of visual data manipulation languages for hybrid data models is HQBE[26], which makes it possible to manipulate relations and structured documents through the visual interface.

Recently, some research groups are studying the visual support for both the data integration and data layout specification in the Web structure construction task. Tiramisu[6] is a well-known system to support the Web structure construction on heterogeneous information sources. The layout specification can be done with visual Web page authoring tools which are familiar to users. However, data integration specification is given by *the site schema*, and full visual support is not attained. InfoWeaver/AQUA[37] provides an integrated visual interface to both specifications. It looks like a typical Web page authoring tool. Its feature is that the system automatically derives the data integration specification and the layout specification from users' instance-based drag-and-drop operations. So the Web structure construction process is totally integrated.

3.4 Other Issues

There are lots of other important issues related to the Web structure construction problem. This section overviews important issues and works which we have not mentioned so far.

There are studies on effective and efficient management frameworks for the Web structure construction. In this article, we have divided this task into the data acquisition and the data layout. Ceri and others[14] identify five perspectives of data-intensive Web sites: sturcture, derivation, navigation, page composition, and presentation. They provide tools to construct Web sites from those perspectives.

The query processing issue is another major concern. Florescu and others[21] attack the problem of when to compute and materialize Web site's contents, and propose optimization techniques exploiting the structure of the Web site. Labrinidis an others[30] compare several materialization policies based on an analytical cost model and actual experiments on an implemented system.

There are several surveys related to the Web structure construction on top of heterogeneous information sources. Kramer[27)] explains technologies to federate the Web and databases, and reports several case studies. Florescu and others[20)] discuss how database concepts can apply to the Web context. It includes modeling and querying the Web, information extraction and integration, and Web site construction and restructuring. Ceri and others[13)] identify ten principles that should be considered when a Web site managing large amounts of data. Fraternali[22)] surveys tools and approaches to develop data-intensive Web applications. This work includes detailed comparison of well-known systems which construct Web structures based on information sources.

As far as data integration is concerned, a lot of research projects on mediation-based systems have been conducted so far[17, 24, 29, 31, 40, 42, 45)]. Domenig and Dittrich[16)] classify mediation-based systems and discuss features of the systems in each class.

4 InfoWeaver/AQUA

This section explains our information integration system named *InfoWeaver* (Figure 7) and its visual facility for the Web structure construction. InfoWeaver follows the mediation approach, and currently accommodates document repositories, the Web, and relational databases as the underlying information sources. It can construct Web structures including HTML, XML, and SMIL[48)] documents on top of the information sources. Its common data model is a hybrid data model which incorporates ADTs for structured documents in nested relational structures. A distinguishing feature of the data model is that it has special operators named *converters* to realize *dynamic data conversion*. This mechanism provides a flexible operational framework for heterogeneous information integration. For example, relational data can be converted to XML documents and vice versa. Another feature is its visual Web structure construction facility, named *AQUA (Amalgamation of QUerying and Authoring)*. It derives both the data integration and data layout specifications by analyzing user's drag-and-drop-based interactive visual operations.

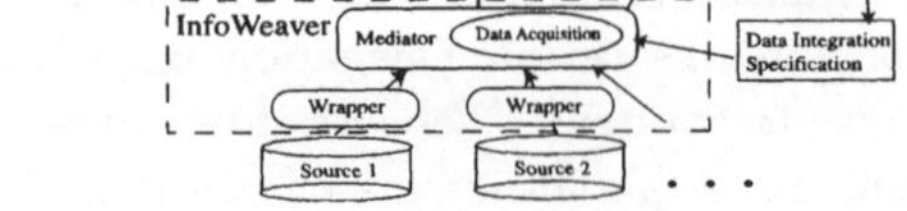

Figure 7: InfoWeaver/AQUA

4.1 Example Application

We consider a relational database and a Web site as information sources. (1) *A baseball game database*: This is a relational database that contains information on baseball games. The relation GAME_VIDEO(<u>Scene_ID</u>, Batter, Pitcher, Contents) records scenes in baseball games and their metadata. The domain of the attribute Contents is an ADT (say, for RealMedia[41)]), named VIDEO type. The meaning of the relation is that each VIDEO value in Contents records a scene specified Scine_ID, where Batter and Pitcher are facing each other. (2) *A baseball information Web site*: This site contains information on baseball teams and players. The Web-site structure is shown in Figure 8(a). The index page contains links to baseball team pages. Each team page contains the team logo (as a reference to a GIF format file) and links to player pages. Each player page contains his information.

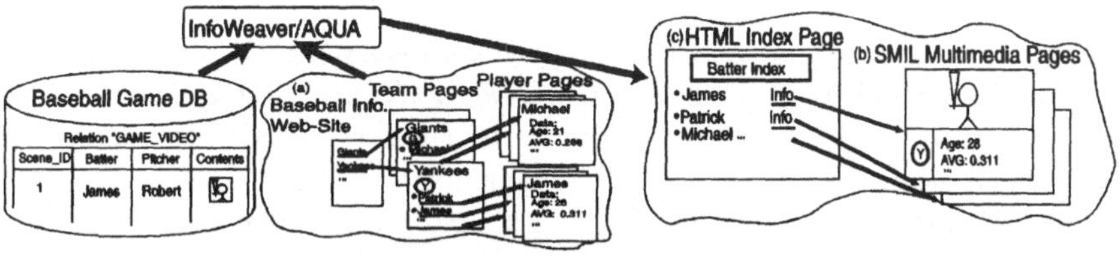

Figure 8: Multimedia Web structure on top of heterogeneous information sources

The requirement here is to create a multimedia Web structure on top of the above information sources. A SMIL Web page is constructed for each player. It is a *multimedia page* (Figure 8(b)), which consists of three different kinds of components: (1) *A sequential rendering of scenes (video objects)* in each of which he is at bat. (2) *The logo of the team he belongs to.* (3) *Text description of his information.*

An index HTML page is also created (Figure 8(c)). It contains an image object ('Batter Index') and links to the players' multimedia pages.

4.2 WebNR/SD Data Model

The core component for integration in InfoWeaver is an integration data model *WebNR/SD*[26, 34)]. WebNR/SD is a hybrid data model that incorporates abstract data types into nested relational structures[23, 43)]. Structured documents such as XML and HTML documents are stored in relations as values of the Structured Document type (SD type) attributes (We call them *SD values*). In Figure 9, attribute B in r_1 and attributes B and D in r_2 are of SD type.

WebNR/SD provides the nested relational algebra operators, and a number of functions associated with SD type to retrieve text elements contained in documents. WebNR/SD has special operators, called *converters*, to dynamically convert structured documents into nested relational structures and vice versa. Moreover,

WebNR/SD has a number of operators for Web data integration. Integration of the Web, structured documents, and relational databases is achieved by combining the above operators. The converters are classified into *Unpack* and *Pack* converters. We explain primitive Unpack and Pack operators here.

Unpack

Unpack (**U**) constructs sub-relations which store text elements originally contained in SD values. We use expressions of *the region algebra*[15] to specify which elements are to be contained in the sub-relations. In this context, *regions* correspond to elements in SD values. A region algebra expression represents a set of regions. For example, region algebra expression "name" returns the set of "name" elements.

Figure 9 gives an example of Unpack, where

$$r_2 := \mathbf{U}_{B \rightarrow C(O,D[name \subset fielders)])\ as\ x}(r_1).$$

It constructs sub-relations for attribute C (with sub-attributes O and D) in relation r_2. Attribute D includes elements which are extracted from SD values in attribute B of r_1, according to the specification $name \subset fielders$. They represent "name" elements which are contained in "fielders" elements.

SD values in attribute B of relation r_2 contain *SD references*, denoted by "`&x.`n`;`." "x" is called a header. The SD references refer to SD values stored in the sub-relation C. In r_2, we call SD values in attribute B *masters*, and those in attribute D *derivatives*.

Pack

Pack (**P**) constructs SD values from sub-relations containing elements. Figure 9 gives an example of Pack, too, where

$$r_1 := \mathbf{P}_{C(O,D)\ as\ x \rightarrow B}(r_2).$$

This Pack restores original SD values in attribute B from sub-relations in attribute C of r_2 and masters in attribute B. The masters are used as templates, and SD references are replaced with elements in derivatives.

With converters, WebNR/SD features the following capabilities: (1) WebNR/SD provides a framework for *symmetrically* and *dynamically* amalgamating relational structures and structured documents. The converters transform SD values into relational structures and vice versa. (2) The converters enable *selective conversion* between SD values and nested relational structures. For example, given a collection of documents with different structures, we can extract common sub-structures and represent them in nested relational structures, ignoring detailed structures.

4.3 AQUA

AQUA[37] is a visual facility for both the data integration and the data layout specification in InfoWeaver. In other systems, the user is usually required to adopt different

r_1:

A	B
1	"<pitchers><name>George</name>...</pitchers> <fielders><name>James</name><name>Patrick</name><fielders>"
2	"<pitchers><name>Robert</name>...</pitchers> <fielders><name>Micheal</name></fielders>"

r_2:

A	B	C	
		O	D
1	"<pitchers><name>George</name>...</pitchers> <fielders>&x.1;&x.2<fielders>"	1	"<name>James</name>"
		2	"<name>Patrick</name>"
2	"<pitchers><name>Robert</name>...</pitchers> <fielders>&x.1;</fielders>"	1	"<name>Micheal</name>"

Figure 9: Unpack and Pack

schemes to query and restructure data and to specify the layout of the result. For example, typical Web application development environments require the user to use SQL for query specification and offer visual tools for designing the layout of the result.

In contrast, InfoWeaver/AQUA provides a visual facility which amalgamates those tasks. The interface looks like just a common authoring tool for HTML and SMIL documents: The user is only required to drag and drop data objects shown in windows into a blank window named the *Canvas.* He can put data objects anywhere he likes, and specify the size of data objects with mouse operations. It is a feature of AQUA that the user can designate an existing data object as an *example.* Then, the data object (the example) serves as the representative of a set of data objects. A drag-and-drop operation of the example is interpreted as manipulation of the set of data objects. Therefore, the object-at-a-time authoring framework and the set-at-a-time data manipulation (querying and restructuring) framework are integrated in a seamless way.

In AQUA, a *DataBox* is used to display a set of data objects stored in the underlying information sources. Figure 10(a)(b)(c) shows example DataBoxes. The DataBox (a) is used to display the relation GAME_VIDEO. In this case, the display unit is a tuple. The user can click the Next and Previous buttons to browse other tuples in the relation. The DataBoxes (b) and (c) are used to display Web pages in the baseball information Web site. In this case, each DataBox is associated with one or more Web pages. The display unit is a Web page. The Web page(s) to be displayed is designated either by specifying a URL or by using a querying facility to gather Web pages. On the screen, there are also *Palettes* (Figure 10(d)) and *Canvas* (Figure 10(e)). Palettes contain decorative objects to be used in Web page design. The Canvas is a blank window into which the user can drag-and-drop data objects from the DataBoxes and Palettes.

A set of objects an example represents is called the *target set.* As a default, the target set is defined as a set of objects each of which appears at the same position on a page as the example object. The user can change the default target set by specifying 'Another' example objects.

When the user specifies multiple examples (and their target sets), he often has to specify *associations* among the target sets. Two types of associations are considered in AQUA. The first one is the *structural association* (S-Association). This occurs according to a structural relationship (relative position) between two examples. For example, if two examples are on the same page, it gives an S-Association that objects from the target sets must be on the same page. The second one is the *value association* (V-Association). This occurs if values of the example objects are the same. When some associations are specified among the target sets, only some combinations of data objects are qualified. Note that an association serves as a kind of join condition.

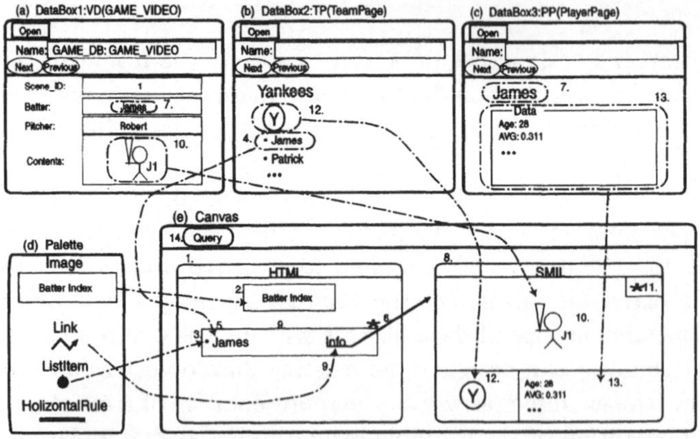

Figure 10: Specification for the example scenario

Figure 10 illustrates the visual specification in AQUA to obtain the Web structure in Section 4.1. The user has to drag-and-drop data objects from the DataBoxes into the Canvas. We explain the operations in the chronological order. (1)First, we open the Canvas and declare the construction of an HTML page. (2)We drag and drop the image 'Batter Index' from the Palette into the Canvas. (3)We drag and drop the 'ListItem' object from the Palette into the Canvas. (4)'James' in TP is specified as an example. The default target set includes those players who appear first in the player list on each team page. Next, 'Patrick' in TP is specified as an 'Another' object. The system uses rules to generalize the relationship between positions of 'James' and 'Patrick,' so that the target set of this example is extended to include all players of all teams. (5)We drag and drop the "James" from TP into the Canvas. (6)We put a repetition mark (*) on the list item object. As a result, all the players' names are listed in this page. Otherwise, a new page is produced for each player. (7)'James' objects in PP and VD are specified as examples. Note that the three target sets of 'James' objects in TP, PP, and VD have V-Association, which specifies equality joins between their target sets. (8)We declare the construction of a SMIL page. (9)We drag and drop the HypertextLink object from the Palette into the Canvas, and connect it to the SMIL page. (10)We specify that '' in VD is an example, and drag and drop it into the Canvas.

(11)We put a repetition mark (*) on the dropped '' object. As a result, the scenes (video objects) of a player are rendered sequentially as one video. Otherwise, a SMIL page is produced for each scene. (12)We specify that '' in TP is an example, and drag and drop it into the Canvas. (13)We specify that the 'Data' section in PP is an example, and drag and drop it into the Canvas. (14)Finally, we press the 'Query' button on the Canvas.

Figure 11 is a screen shot of the prototype system where the user is specifying the example scenario. These operations are analyzed in AQUA, and the data integration requests are issued to the InfoWeaver mediator. AQUA processes the integration result and generates the specified Web structure.

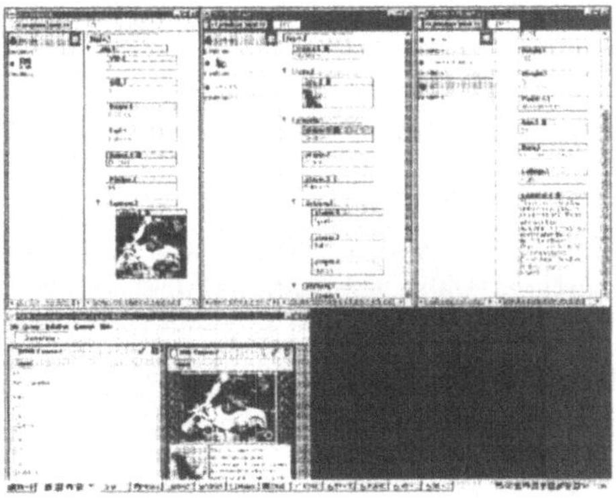

Figure 11: A screen shot of the prototype system

5 Conclusion

With the broad acceptance of the World Wide Web, the Web has been widely used for publishing and disseminating information stored in various information sources. Therefore, the construction of Web structures on top of the heterogeneous information sources has become an important research issue. In this article, we surveyed the current approaches to the problem in the first part, and showed our approach in the second part. The current approaches to the problem can be classified into the three approaches. Among them, we mainly explored the mediation approach, and gave an overview of the technology related to this from a number of important design viewpoints. In the second part, we introduced our approach named InfoWeaver/AQUA. InfoWeaver provides the mediation function for data integration. The common data model WebNR/SD is a hybrid data model based on nested relations and ADT values representing structured documents. AQUA is a drag-and-drop-based interactive visual facility for the Web structure construction in InfoWeaver, and supports the data integration and data layout specifications in a seamless manner. The main function of InfoWeaver/AQUA has been implemented and is operational, and some extended visual facilities are under development.

Acknowledgments

The authors are grateful to the contribution of many members of their research group to the development of InfoWeaver/AQUA. They also thank Nippon Television Network Corporation for providing sample video data. This work has been supported in part by the Grant-in-aid for Scientific Research from Japan Society for the Promotion of Science.

Bibliography

1) S. Abiteboul, Querying semi-structured data. *Proc. ICDT'97*, 1997, pp. 1-18.

2) S. Abiteboul, P. Buneman and D. Suciu, *Data on the Web*, Morgan Kaufmann, 1999.

3) S. Abiteboul, D. Quass, J. McHugh, J. Widom and J. L. Wiener, The Lorel Query Language for Semistructured Data. *International Journal on Digital Libraries*, Vol. 1, No. 1, 1997, pp.68-88.

4) S. Adali, K. S. Candan, Y. Papakonstantinou and V. S. Subrahmanian, Query Caching and Optimization in Distrubuted Madiator Systems. *Proc. ACM SIGMOD'96*, May 1996, pp.137-148.

5) Allaire Corporation, Allaire Homepage. http://www.allaire.com/.

6) C. R. Anderson, A. Y. Levy and D. S. Weld, Declarative Web-Site Management with Tiramisu. *Proc. ACM SIGMOD Workshop on the Web and Databases (WebDB'99)*, 1999.

7) G. O. Arocena and A. O. Mendelzon, WebOQL: Restructuring Documents, Databases, and Webs. *Proc. ICDE'98*, pp.24-33, Feb. 1998.

8) P. Atzeni, G. Mecca and P. Merialdo, To Weave the Web. *Proc. VLDB'97*, Aug. 1997, pp. 206-215.

9) C. K. Baru, A. Gupta, B. Ludäscher, R. Marciano, Y. Papakonstantinou, P. Velikhov and V. Chu, XML-Based Information Mediation with MIX. *Proc. ACM SIGMOD'99*, May 1999, pp.597-599.

10) L. Bouganim, T. Chan-Sine-Ying, T. Dang-Ngoc, J. Darroux, G. Gardarin and F. Sha, Miro Web: Integrating Multiple Data Sources through Semistructured Data Types. *Proc. VLDB'99*, 1999, pp. 750-753.

11) P. Buneman, Semistructured data. *Proc. ACM PODS'97*, May 1997, pp. 117-121.

12) S. Ceri, S. Comai, E. Damiani, P. Fraternali, S. Paraboschi and L. Tanca, XML-GL: A Graphical Language for Querying and Restructuring XML Documents. *WWW8 / Computer Networks*, Vol. 31, No. 11-16, 1999, pp. 1171-1187.

13) S. Ceri, P. Fraternali and S. Paraboschi, Design Principles for Data-Intensive Web Sites. *ACM SIGMOD Record*, vol. 28, no. 1, Mar. 1999, pp. 84-49.

14) S. Ceri, P. Fraternali and S. Paraboschi, Data-Driven, One-To-One Web Site Generation for Data-Intensive Applications. *Proc. VLDB'99*, 1999, pp. 615-626.

15) M. P. Consens and T. Milo. Algebras for Querying Text Regions. *Proc. ACM PODS'95*, May 1995, pp. 11-22.

16) R. Domenig and K. R. Dittrich, An Overview and Classification of Mediated Query Systems. *ACM SIGMOD Record*,Vol. 28, No. 3, 1999, pp. 63-72.

17) A. Elmagarmid, M. Rusinkiewicz and A. Sheth (eds.), *Management of Heterogeneous and Autonomous Database Systems,* Morgan Kaufmann, 1999.

18) M. Fernandez, D. Florescu, J. Kang, A. Levy and D. Suciu, STRUDEL: A Web-site Management System. *Proc. ACM SIGMOD'97*, Tucson, May 1997, pp. 549-552.

19) M. F. Fernandez, D. Florescu, J. Kang, A. Levy and D. Suciu, Catching the Boat with Strudel: Experiences with a Web-Site Management System. *Proc. ACM SIGMOD'98*, 1998, pp. 414-425.

20) D. Florescu, A. Levy and A. Mendelzon, Database Techniques for the World-Wide Web: A Survey. *ACM SIGMOD Recrd*, vol. 27, no. 3, 1998, pp. 59-74.

21) D. Florescu, A. Levy, D. Suciu and K. Yagoub, Optimization of Run-time Management of Data Intensive Web Sites. *Proc. VLDB'99*, 1999, pp. 627-638.

22) P. Fraternali, Tools and approaches for developing data-intensive Web applications: a survey. *ACM Computing Surveys*, vol. 31, no. 3, , Sep. 1999 pp. 227 - 263.
pp. 34-41, May 1997.

23) P. C. Fischer and S. J. Thomas, Operators for Non-First-Normal-Form Relations. *Proc. IEEE COMPSAC83*, Chicago, 1983, pp. 464-475.

24) M. R. Genesereth, A. M. Keller and O. M. Duschka, Infomaster: An Information Integration System. *Proc. ACM SIGMOD'97*, Tucson, May 1997, pp. 539-542.

25) R. Goldman and J. Widom, DataGuides: Enabling Query Formulation and Optimization in Semistructured Databases. *Proc. VLDB '97*, 1997, pp.436-445.

26) H. Kitagawa, A. Morishima and H. Mizuguchi, Integration of Heterogeneous Information Sources in InfoWeaver. *Advances in Databases and Multimedia for the New Century - A Swiss/Japanese Perspective -*, Advanced Database Research and Development Series, Vol. 10, World Scientific, 2000, pp. 124-137.

27) R. Kramer, Databases on the Web: Technologies for Federation Architectures and Case studies. *Proc. ACM SIGMOD'97*, 1997, pp. 503-506.

28) L. Lakshmannan, F. Sadri and I. Subramanian, A Declarative Language for Querying and Restructuring the Web. *Proc. 6th International Workshop on Research Issues in Data Eng. (RIDE'96)*, Feb. 1996.

29) Y. Lee, L. Liu and Calton Pu, Towards Interoperable Heterogeneous Information Systems: An Experiment Using the DIOM Approach. *Proc. ACM SAC'97*, Feb. 1997, pp.112-114.

30) A. Labrinidis and N. Roussopoulos, WebView materialization. *Proc. ACM SIGMOD'00*, 2000.

31) A. Y. Levy, A. Rajaraman and J. J. Ordille, Querying Heterogeneous Information Sources Using Source Descriptions. *Proc. VLDB'96*, Sep. 1996, pp.251-262.

32) Macromedia, Inc., Macromedia Homepage. http://www.macromedia.com/.

33) Microsoft Corporation, Microsoft Homepage. http://www.microsoft.com/.

34) A. Morishima and H. Kitagawa, Integrated Querying and Restructuring of the World Wide Web and Databases. *Proc. International Symposium on Digital Media Information Base (DMIB'97)*, Nov. 1997, pp. 261-271.

35) A. Morishima and H. Kitagawa, InfoWeaver: Dynamic and Tailor-Made Integration of Structured Documents, Web and Databases. *Proc. ACM Digital Libraries '99*, Aug. 1999, pp. 235-236.

36) A. Morishima, H. Kitagawa, H. Mizuguchi and S. Koizumi, Dynamic Creation of Multimedia Web Views on Heterogeneous Information Sources. *Proc. 33rd Hawaii International Conference on System Sciences (HICSS-33)*, Jan. 2000.

37) A. Morishima, S. Koizumi and H. Kitagawa, Drag and Drop: Amalgamation of Authoring, Querying and Restructuring for Multimetia View Construction. *Proc. 5th IFIP 2.6 Working Conference on Visual Database Systems*, May 2000, pp. 257-276.

38) Netscape Communications Corporation, Netscape Homepage. http://www.netscape .com/.

39) T. Nguyen and V. Srinivasan, Accessing Relational Databases from the World Wide Web. *Proc. ACM SIGMOD'96*, 1996, pp. 529-540.

40) Y. Papakonstantinou, H. Garcia-Molina and J. Widom, Object Exchange Across Heterogeneous Information Sources. *Proc. ICDE'95*, Mar. 1995, pp. 251-260.

41) RealNetworks, Inc., RealNetworks Home Page. http://www.real.com/.

42) M. T. Roth and P. M. Shwarz, Don't Scrap It, Wrap It! A Wrapper Architecture for Legacy Data Sources. *Proc. VLDB'97*, Sep. 1997, pp. 266-275.

43) H.-J. Schek and M. H. Scholl, The Relational Model with Relation-valued Attributes. *Information Systems*, Vol. 11, No. 2, 1986, pp.137-147.

44) Sun Microsystems, Inc., Java Servlet API, http://www.java.sun.com/products/servlet/.

45) A. Tomasic, R. Amouroux, P. Bonnet, O. Kapitskaia, H. Naacke and L. Raschid, The Distributed Information Search Component (Disco) and the World Wide Web. *Proc. ACM SIGMOD'97*, May 1997, pp.546-548.

46) The Apache/Perl Integration Project, Embperl - Embed Perl in Your HTML Documents. http://perl.apache.org/embperl.

47) G. Wiederhold, Mediators in the Architecture of Future Information Systems. *IEEE Computer*, Vol. 25, No. 3, 1992, pp. 38-49.

48) W3C. Synchronized Multimedia Integration Language (SMIL) 1.0 Specification. W3C Recommendation, http://www.w3.org/TR/REC-smil.

49) W3C. XML-QL: A Query Language for XML. W3C note, http://www.w3.org/TR/.

50) M. M. Zloof, Query by Example: A Data Base Language. *IBM Systems Journal*, Vol. 16, No. 4, 1977, pp.324–343.

13

Parallel Execution of SQL Based Association Rule Mining

Masaru Kitsuregawa
Iko Pramudiono
Institute of Industrial Science, The University of Tokyo

Takeshi Yoshizawa
IBM Japan Co.,Ltd

Takayuki Tamura
Mitsubishi Electric

ABSTRACT

Association rule mining over large scale databases has been recognized as powerful tool to extract hidden precious information from those databases. However in most cases, user has to pull out data from database and relies on external specialized program to perform the mining. Here we present our examination on association rule mining using native SQL on parallel platforms such as experimental PC cluster and also commercial parallel RDBMS. The integration on RDB framework offers the portability, and ease-of-maintenance. We show that parallelism is the key to achieve sufficient performance.

1 Introduction

In the business world data mining over data warehouse has become a crucial weapon to gain competitive edge against competitors. Those organizations have accumulated large amount of transaction data by mean of data collection tools such as POS and they want to extract value added information such as unknown buying patterns from that large databases. This demand has fueled the growing popularity of data mining.

One method of data mining is finding association rule that is a rule which implies certain association relationship such as "occur together" or "one implies the other" among a set of objects [1)]. This mining that is also known as "basket data analysis" retrieves information like "90% of the customers who buy A and B also buy C" from transaction data.

Association rule mining is a kind of mining that is known as CPU power demanding application. This fact has driven many initial researches in data mining to develop new efficient mining methods such as Apriori [2] and its improvements [9][3]. Some algorithms are already available as commercial packages. Most of them assumes the data is stored in flat file system. However in most case, the data is managed by RDBMS. Thus one has to export the data from database and perform the data mining with specialized software outside the database. Some softwares also provide data access to database using cursor interface [7].

However RDBMS has sophisticated query processing capability by means of standard language SQL. Therefore there are some efforts recently to perform data mining using relational database system which offer advantages such as seamless integration with existing system and high portability. Some methods examined ranging from directly using SQL to some extensions like user defined function (UDF) [11]. Some efforts have been conducted to couple RDBMS more tightly with association rule mining system. For example DMQL [5] and M-SQL [8] proposed some SQL standard extensions to handle mining operators.

Pure SQL-92 approach is interesting since SQL-92 is standard supported by most database system which means it offers the highest level of portability and flexibility. Unfortunately SQL approach is reported to have drawback in performance.

We proposed large-scale PC cluster as cost effective platform for data intensive applications such as data mining using parallel RDBMS, which offers the advantages of the integration without sacrificing the performance [13].

There is a tradeoff between performance and portability. Performance is not necessarily sufficiently high but seamless integration with existing RDBMS would be considerably advantageous. Since RDB is already very popular, the feasibility of association rule mining can be explored using query of standard SQL instead of purchasing expensive mining software. In addition, parallel RDB is now also widely accepted. We showed that paralleling the SQL execution of modified SETM query on PC cluster can offer the same performance as those Apriori based native programs with 4 nodes. Since most organizations have a lot of PCs, which are not fully utilized. We are able to exploit such resources to enhance the performance significantly.

On the other hand recently most major commercial database systems have included capabilities to support parallelization although no report available about how the parallelization affects the performance of complex query required by association rule mining. This fact motivated us to examine how efficiently SQL based association rule mining can be parallelized and speeded up using commercial parallel database system (IBM DB2 UDB EEE). We propose two techniques to enhance association rule mining query based on SETM [3]. And we have also compared the performance with commercial mining tool (IBM Intelligent Miner). Our performance evaluation shows that we can achieve comparable performance with commercial mining tool using only 4 nodes.

This paper is composed with 6 sections. In second section we will briefly explain association rule mining. The SQL queries to mine association rule will be described in third section. The evaluation on experimental PC cluster will given in fourth section. While fifth section shows how currently available parallel commercial RDBMS performs with the queries.

2 Association Rule Mining

A typical example of association rule is "if a customer buys A and B then 90% of this kind of customers buy also C" . Here 90% is called the $confidence$ of the rule. Another measure of a rule is called the *support* of the rule.

Transactions in a retail database usually consist of an identifier and a set of items or itemset. $\{A, B, C\}$ in above example is an itemset. An association rule is an implication of the form $X \Longrightarrow Y$ where X and Y are itemsets. An itemset X has support s if s% of transactions contain that itemset, here we denote $s = support(X)$. The support of the rule $X \Longrightarrow Y$ is $support(X \cup Y)$. The $confidence$ of that rule can be written as the ratio $support(X \cup Y)/support(X)$.

The problem of mining association rules is to find all the rules that satisfy a user-specified minimum support and minimum confidence, which can be decomposed into two subproblems:

1. Find all combinations of items, called large itemsets, whose support is greater than minimum support.

2. Use the large itemsets to generate the rules.

Since the first step consumes most of processing time, development of mining algorithms has been concentrated on this step.

3 Association Rule Mining Algorithms on SQL

Most of the algorithms developed to mine association rule was intended to pursuit effectiveness so somehow they neglect integration with existing system. Some exception such as SETM[4)] reported SQL expression of association rule mining.

The ability to do data mining directly on RDBMS using SQL provides many benefits among others :

1. Small implementation cost

   Since we can use SQL available on all RDBMS to do data mining, we don't have to buy expensive data mining software separately. Organizations that are still considering the introduction of full-scale data mining can easily set up experimental dataset and test the efficiency of data mining applications using existing RDBMS capability.

2. SQL as standard language

   The popularity of SQL as standard language for manipulating database may shorten the time required to implement data mining using SQL. Furthermore the query can be easily enhanced and customized according to the needs since the query is well defined based on simple concepts.

3. Integration with RDBMS

   Seamless integration with RDBMS reduces cost of maintenance and maximize the portability since the difference between platforms can be absorbed by the RDBMS. In addition, mature technologies used in RDBMS such as query optimizations, parallelization, indexes, checkpoints so on are available at no extra cost.

In our evaluation we employ a modified version of SETM. For the implementation on commercial parallel RDBMS, we could utilyze some other techniques to enhance the query. Here we introduce using view and subquery to reduce disk I/O.

Recently pure SQL implementation of the well known Apriori algorithm [2] has been reported but the performance is far behind its object oriented SQL extensions or other more loosely integrated approachs [11].

Sarawagi et.al. extended the query to mine generalized association rule with taxonomy [10]. In addition they also extended the query further to handle sequential pattern as well. Analysis of execution plan has given some hints to improve performance of the Apriori based query [12].

4 Representation of Transaction Data

The transaction data can be representated in relational database using first normalization such as ilustrated in Table 1. The schema for the table is $SALES(TID, item)$ where TID represents transaction ID and $item$ represents item code or item name. For each customer transaction that takes place, tuples corresponding to every items are inserted into SALES.

4.1 Modified SETM

The first SQL query available to perform flat association rule is called SETM [4].

In our experiments we employed ordinary standard SQL query that is similar to SETM algorithm. We modified the query to enable hash join execution. It is shown in figure 1.

In the first pass we simply gather the count of each item. Items that satisfy the minimum support inserted into large itemsets table C_1 that takes form(item, item count). Then transaction data that match large itemsets stored in R_1.

Transaction *SALES*

TID	Items
100	1,3,4
200	2,3,5
300	1,2,3,5
400	2,5

⇒

Relation *SALES*

TID	Items
100	1
100	3
100	4
200	2
200	3
200	5
300	1
300	2
300	3
300	5
400	2
400	5

Table 1: Representation of transaction data in relational database

```
CREATE TABLE SALES (id int, item int);

- PASS 1

CREATE TABLE C_1 (item_1 int, cnt int);
CREATE TABLE R_1 (id int, item_1 int);

INSERT INTO C_1
    SELECT    item AS item_1, COUNT(*)
    FROM      SALES
    GROUP BY item
    HAVING COUNT(*) >= :min_support;

INSERT INTO R_1
    SELECT    p.id, p.item AS item_1
    FROM      SALES p, C_1 c
    WHERE     p.item = c.item_1;

- PASS k
CREATE TABLE RTMP_k (id int, item_1 int,
    item_2 int, ... , item_k int)
CREATE TABLE C_k       (item_1 int,
    item_2 int, ... , item_k int, cnt int)
CREATE TABLE R_k       (id int, item_1 int,
    item_2 int, ... , item_k int)

INSERT INTO RTMP_k
    SELECT    p.id, p.item_1, p.item_2, ... ,
              p.item_k-1, q.item_k-1
    FROM      R_k-1 p, R_k-1 q
    WHERE     p.id       = q.id
      AND     p.item_1 = q.item_1
      AND     p.item_2 = q.item_2
                     .
                     .
                     .
      AND     p.item_k-2 = q.item_k-2
      AND     p.item_k-1 ¡ q.item_k-1;

INSERT INTO C_k
    SELECT    item_1, item_2, ..., item_k,
              COUNT(*)
    FROM      RTMP_k
    GROUP BY item_1, item_2, ..., item_k
    HAVING COUNT(*) >= :min_support;

INSERT INTO R_k
    SELECT    p.id, p.item_1, p.item_2, ...,
              p.item_k
    FROM      RTMP_k p, C_k c
    WHERE     p.item_1 = c.item_1
      AND     p.item_2 = c.item_2
                     .
                     .
                     .
      AND     p.item_k = c.item_k;

DROP TABLE R_k-1;
DROP TABLE RTMP_k;
```

Figure 1: SQL query to mine association rule

In other passes for example pass k, we first generate all lexicographically ordered candidate itemsets of length k into table RTMP_k by joining $k-1$ length transaction data. Then we generate the count for those itemsets that meet minimum support and included them into large itemset table C_k. Finally transaction data R_k of length k generated by matching items in candidate itemset table RTMP_k with items in large itemsets.

4.2 Enhanced SETM query using view materialize technique

SETM has to materialize its temporary tables namely R_k and RTMP_k. Those temporary tables are only required in the next pass and they are not needed for generating the rules. In fact, those tables can be deleted after execution of its subsequent pass. Based on this observation we could avoid materialization cost of those temporary tables by replacing the table creation with view.

4.3 Enhanced SETM query using subquery technique

We expect significant performance improvement with utilization of view, however view still requires time to access the system catalog and are holding locks to the system catalog table during creating views so we further use subquery instead of temporary tables. Therefore we embed the generation of item combinations into the query to generate large itemsets.

4.4 Apriori SQL

Sarawagi et.al. proposed SQL query to mine association rule that is based on Apriori algorithm [11)]. We omit the detail of the query due to space limitation.

The query differs from SETM since it first generate the candidate of large itemsets before support counting process. The candidate table at pass k is generated by joining two copies of large itemsets with length $(k-1)$ from previous pass. The join result which is a set of k-itemsets is further pruned using the subset pruning strategy that all subsets of a large itemset should be frequent.

They also propose some methods to do support counting such as K-way join, 3-way join, 2Groupby and Subquery. The Subquery method is reported to have best performance.

However they also pointed out that the performance of pure SQL-92 implementations is far behind their counterparts such as native programs or OODB based queries that utilyze user defined function(UDF).

4.5 Set-oriented Apriori

After investigation of several execution plans of Apriori SQL, some modifications are proposed to improve the performance [12)]. The modifications are : pruning non-frequent items from the transaction database after the first pass, elimination of candidate generation in second pass since the number of candidates are too many to materialize and reusing the item combinations from previous pass in the similar way as the SETM.

The modified query is then called Set-oriented Apriori. However we use the simpler modified SETM in this evaluation since we found that the performance doesn't differ too much for our dataset.

5 Performance Evaluation on PC Cluster

At present, parallel SQL is running on expensive massively parallel machines but not in the future. Instead it will run on inexpensive PC cluster system or WS cluster system. Thus we believe that SQL implementation based on sophisticated optimization would be one of reasonable approaches.

5.1 Parallel Execution Environment

The experiment is conducted on a PC cluster developed at Institute of Industrial Science, The University of Tokyo. This pilot system consists of one hundred commodity PCs connected by ATM network named NEDO-100. We have also developed DBKernel database server for query processing on this system. Each PC has Intel Pentium Pro 200MHz CPU, 4.3GB SCSI hard disk and 64 MB RAM.

The performance evaluation using TPC-D benchmark on 100 nodes cluster is reported[13)]. The results showed it can achieve significantly higher performance especially for join intensive query such as query 9 compared to the current commercially available high end systems.

5.2 Dataset

We use synthetic transaction data generated with program described in Apriori algorithm paper[2)] for experiment. The parameters used are : number of transactions 200000, average transaction length 10 and number of items 2000. Transaction data is partitioned uniformly correspond to transaction ID among processing nodes' local hard disk.

5.3 Results

The execution times for several minimum support is shown in figure 2(left). The result is surprisingly well compared even with directly coded Apriori-based C program on single processing node. On average, we can achieve the same level of execution time by parallelizing SQL based mining with around 4 processing nodes. The speedup ratio shown in figure 2(right) is also reasonably good, although the speedup seems to be saturated as the number of processing nodes increased. As the size of the dataset asssigned to each node is getting smaller, processing overhead and also synchronizing cost that depends on the number of nodes cancel the gain.

Figure 3(left) shows the time percentage for each pass when the minimum support is 0.5%. Eight passes are necessary to process entire transaction database. It is well known that in most cases the second pass generates huge amount of candidate

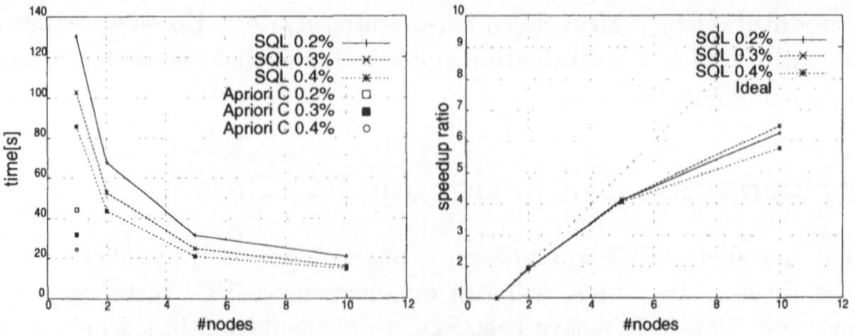

Figure 2: Execution time(left) Speedup ratio(right)

itemsets thus it is the most time consuming phase. Figure 3(right) shows the speedup ratio for each pass. The later passes, the smaller candidate itemsets. Thus non-negligible parallelization overhead become dominant especially in passes later than five. Depending on the size of candidate itemsets, we could change the degree of parallelization. That is, we should reduce the number of nodes on later passes. Such extensions will need further investigations.

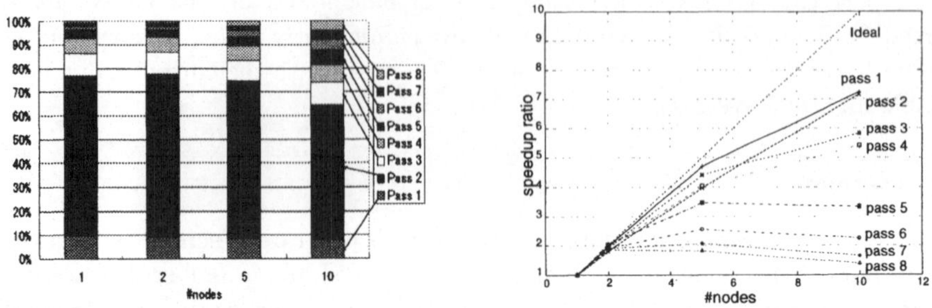

Figure 3: Pass analysis (minimum support 5%). Contribution of each pass in execution time(left) Speedup ratio of each pass(right)

5.4 Execution Behaviour

Original SETM algorithm assumes execution using sort-merge join[4)]. Although they have showed that sort-merge join is better than nested loop joi n with indexes, sort process is hardly parallelable. Inside database server on our system, relational joins are executed using hash joins and tables are partitioned over nodes by hashing. As the result, parallelization efficiency is much improved. This approach is very effective for large scale data mining.

The DB Kernel allows user to freely custom the execution plan of any query. We have made the execution plan to accommodate the hash join while suppress the communication among nodes to achieve better speedup ratio.

Inside execution plan for this query, join to generate candidate itemsets is executed at each node independently. And then after counting itemsets locally, nodes exchange the local counts by applying hash function on the itemsets in order to determine overall support count of each itemset. Another data exchange occurs when large itemsets are distributed to every nodes. Finally each node independently execute join again to create the transaction data.

Execution plan for SQL statement of pass k which is shown in Figure 1 is here:
Pass k:

1. Build hash table from R_k-1.Probe the hash table using table R_k-1 itself, insert tuples that satisfy select conditions into table RTMP_k.

2. Count support in the local node.

3. In order to count global support, distribute the count result to processing nodes by applying hash function on items.

4. Count global support for each item.

5. Broadcast items that satisfy the minimum support to every nodes and insert to table C_k.

6. Build hash table from C_k.Probe hash table using transaction data, insert tuples that satisfy select conditions into table R_k.

The detail of the execution plan is shown in Figure 4. Figure 5 give graphical description of the execution plan.

We have also examined some other alternatives such as instead of distributing the result of support counting in each node to obtain global item support we distribute the tuples each pass by applying hash function on items before support counting, therefore all support counting can be performed in one step. But preliminary test for this plan has shown poor performance. In order to suppress the communication cost this plan requires most of items remained at sender node. Our preliminary test showed that more tuples are redistributed so this plan is impractical unless the query processor is equipped with mechanism to detect the distribution of items in each node.

This kind of query utilyze the so called self join extensively. The self join is a join between the same table. Proper optimization of self join will definitely make the execution faster.

DBKernel provides facility to trace the usage of resource during execution as shown in figure 6. Second pass occupies interval from 3s until 23.5s.

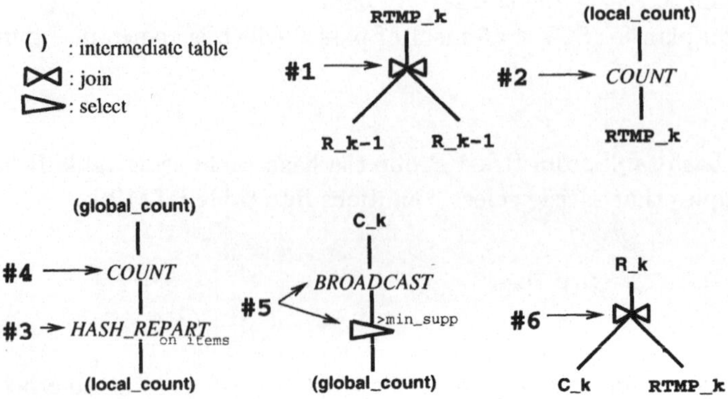

Figure 4: Execution plan

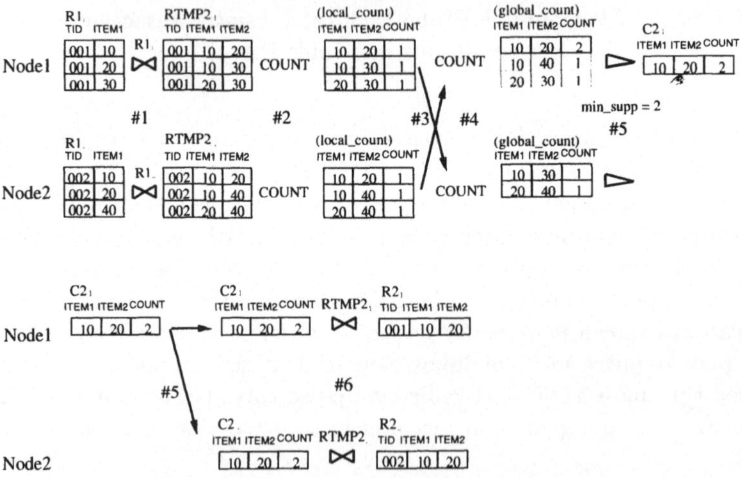

Figure 5: Illustration of the execution plan

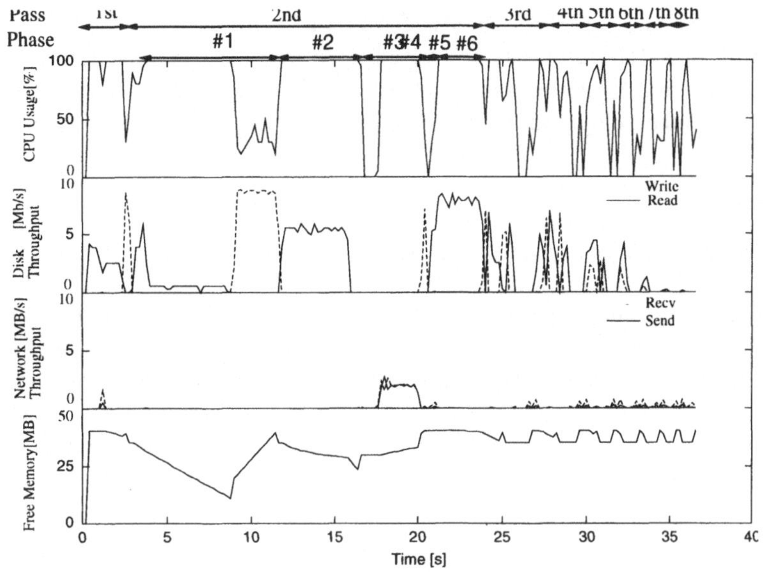

Figure 6: Execution trace(200k 0.5% 5 nodes)

In candidate generation, occured during interval from 3.5s to 12s and also shown in figure as phase #1, heavy probe operation during join in the first half of this phase resulting in 100% CPU load and low disk reading throughput. In later part of this phase disk I/O bound occured when the result of the join stored in disk. CPU bound is also observed in other phase involving hash table probing such as candidate matching (phase #6, 21.5s - 23.5s) and during global support gathering (phase #2, 12s - 16.5s) which employs hash table updating for aggregation. However significant network throughput dominates the global support gathering (phase #3 and #4, 16.5s - 21.5s) when processing nodes exchanging their local support counts.

6 Performance Evaluation using Commercial Parallel RDBMS

Since the size of databases and the amount of required processing power has increased incredibly, parallel processing ability has become a must for commercial RDBMS. It is interesting to know whether currently available technology can achieve sufficient performance when handling complex query such as association rule mining.

6.1 Parallel Execution Environment

In our experiment we employed commercial Parallel RDBMS: IBM DB2 UDB EEE version 6.1 on IBM UNIX Parallel Server System: IBM RS/6000 SP. 12 nodes make this system and using shared nothing architecture. Each node has POWER2 77Mhz

CPU, 4.4GB SCSI hard disk, 256MB RAM and connected by High Performance Switch HPS with 100MB/s network speed. We also used commercial data mining tool IBM Intelligent Miner on single node of RS/6000 SP for performance comparison with the SQL based data mining.

6.2 Dataset

We show execution time of association rule mining with several minimum supports in Figure 7. The data used here is synthetic transaction data generated with program described in Apriori algorithm [2)] to show that we can handle larger data with parallel RDBMS. The number of transactions here is 200000, average transaction length 10 and number of items 2000. Transaction data is partitioned uniformly by hashing algorithm corresponds to transaction ID among processing nodes' local hard disks. We also show the result of commercial data mining program Intelligent Miner from IBM on single node for reference.

6.3 Results

Figure 7 shows the execution time on each degree of parallelization. On average, we can derive that View and Subquery SQL is about 6 times faster than SETM SQL regardless of the number of nodes. The result is also compared with the execution time of Intelligent Miner on single processing node. It is true that Intelligent Miner on single node with transaction data stored in flat file is much faster than the SQL queries. However, the View and Subquery SQL are 50% faster than Intelligent Miner on single node if the transaction data have to be read from RDBMS. We exemplified that we can achieve comparable performance of Intelligent Miner on single node with flat file by activating only 4 nodes when we used View and Subquery SQL. The result gives evidence for the effectiveness of parallelization of SQL query to mine association rule.

The speedup ratio is shown in Figure 8. This is also reasonably good, especially View and Subquery SQL are not being saturated as the number of processing nodes increased. That means they can be parallelized well. The execution is 11 times faster with 12 nodes. In parallel environment, network potentially becomes botleneck which degrades the speed-up ratio. However our experiments suggest that association rule mining using variants of SETM is mostly CPU bound and network I/O is negligible.

Here we give thorough comparison and analysis on the three variations of SQL query described before. The performance evaluation is done on 12 nodes. The mining is two passes long. It is well known that in most cases the second pass generates huge amount of candidate itemsets thus it is the most time consuming phase[4][5]. Our results are very much alike. Almost over 80% of execution time belongs to PASS2 in all three SQL queries. Obviously View and Subquery SQL complete their first and second passes faster than SETM SQL query. We have recorded the execution traces of the three SQL in each PASS. The decomposition of execution time is analysed as shown Figure 9 (PASS2) respectively. Comparing the elapsed time with the cpu time

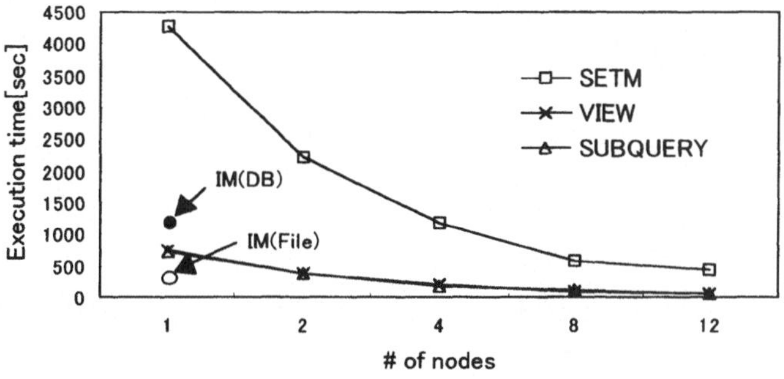

Figure 7: Execution time on parallel database environment

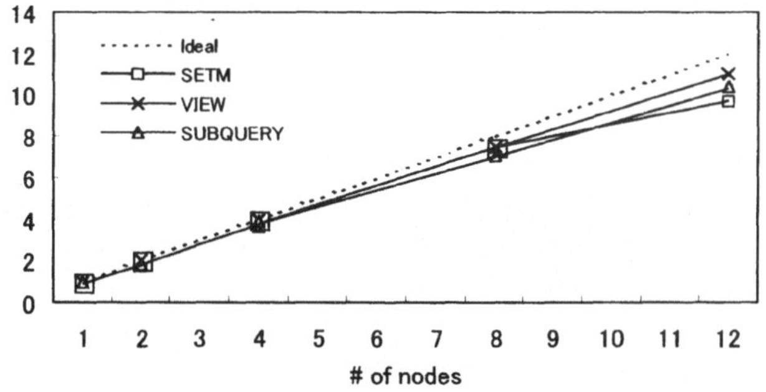

Figure 8: Speedup ratio in parallel database environment

at Figure 9, we find that both are close for View SQL and Subquery SQL. This means these SQL's are cpu bound, while SETM SQL is not cpu bound. Most of execution time of SETM query is dominated by disk write time for creating temporary table such as R_k and RTMP_k. We can also see that sort time is almost equal for all three SQL's, which represents the cost of group by aggregation. In PASS2, SETM reuses item combinations in temporary table R1 on the secondary storage that is generated in PASS1. We replace it with view or subquery. Then data is transferred directly through memory from PASS1 to PASS2. Figure 9 indicates that PASS2 of those modified SQL queries only read data from buffer pool. Thus the disk write time of View SQL and Subquery SQL is almost negligible, although it is dominant for SETM SQL. This analysis clarifies the problem of SETM and how to cost can be reduced for View and Subquery SQLs, which is the key to the performance improvement.

7 Summary and Conclusion

The integration of data management and data processing on RDBMS platform has several potential advantages. The standard such as SQL enables portability among

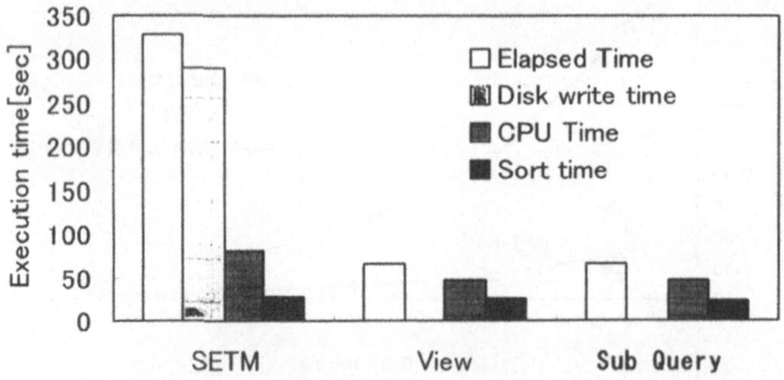

Figure 9: Decomposition of execution time of PASS2 for three types of SQL queries

different platforms and flexible enhancements. Established technologies in RDBMS such as parallelization, memory management, query processing and failure recovery come without extra cost.

The performance degradation can be compensated with sufficient parallelism that we showed in this paper. In fact elimination of the operation to move data from database to specialized data mining software not only reduce maintenance cost but also improve the processing time in overall.

PC cluster is a prospective platform for parallel RDBMS with its high cost-performance. The redundant processing power of PCs in offices could be exploited to perform complex queries such as needed by data mining applications.

Through real implementation, we have confirmed that parallelization using 4 processing nodes for association rule mining can beat the performance of directly coded C implementation. We do not have to buy or write special data mining application codes, SQL query for association rule mining is extremely easy to implement. It is also very flexible, we have extended the modified SETM SQL query to handle generalized association rule mining with taxonomy [6)].

There remains lots of further investigations. Since our system has 100 nodes, we could handle larger transaction database. In such experiments, data skew become a problem. Skew handling is one of the interesting research issue for parallel processing.

We have also examined the parallelization of SQL query to mine association rule on commercial RDBMS (IBM DB2 UDB EEE). We showed that good speedup ratio can be achieved, that means it is parallelized well. We also examined two variations of SETM SQL queries to improve performance, which reduce I/O cost by using View materialize or Subquery technique.

We have compared the parallel implementation of SQL based association rule mining with commercial data mining tool (IBM Intelligent Miner). Through real implementation, we have showed that our improved SETM query using View or Subquery can beat the performance of specialized tool with only 4 nodes while original SETM query would need more than 24 processing nodes to achieve the same performance.

Bibliography

1) R. Agrawal, T. Imielinski, A. Swami. Mining Association Rules between Sets of Items in Large Databases. In Proc. of the ACM SIGMOD Conference on Management of Data, 1993.

2) R. Agrawal, R. Srikant. Fast Algorithms for Mining Association Rules. In Proc. of the VLDB Conference, 1994.

3) S. Brin, R. Motwani, J.D. Ullman, S. Tsur. Dynamic Itemset Counting and Implication Rules for Market Basket Data. In Proc. of the ACM SIGMOD Conference on Management of Data, 1997.

4) M. Houtsma, A. Swami. Set- oriented Mining of Association Rules. In Proc. of International Conference on Data Engineering, 1995.

5) J. Han, Y. Fu, K. Koperski, W. Wang, O. Zaiane. DMQL: A Data Mining Query Language for Relational Databases. In Proc. of the ACM SIGMOD Conference on Management of Data, 1996.

6) I. Pramudiono, T. Shintani, T. Tamura, M. Kitsuregawa. Parallel SQL Based Association Rule Mining on Large Scale PC Cluster: Performance Comparison with Directly Coded C Implementation. In Proc. of First International Conference on Data Warehousing and Knowledege Discovery (DAWAK), 1999.

7) T. Imielinski, H. Manilla. A database perspective on knowledge discovery. Communication of the ACM, 39(11):58-64, Nov 1996.

8) T. Imielinski, A. Virmani, A. Abdulghani. Discovery Board Application Programming Interface and Query Language for Database Mining. In Proc. of Conference of Knowledge Discovery and Data Mining (KDD), 1996.

9) J. S. Park, M. S. Chen, and P. S. Yu. An Effective Hash-Based Alogorithm for Mining Association Rules. In Proc. of the ACM SIGMOD Conference on Management of Data, 1995.

10) S. Sarawagi, S. Thomas. Mining Generalized Association Rules and Sequential Patterns Using SQL Queries. In Proc. of Conference of Knowledge Discovery and Data Mining (KDD), 1998.

11) S. Sarawagi, S. Thomas, R. Agrawal. Integrating Association Rule Mining with Relational Database Systems: Alternatives and Implications. In Proc. of the ACM SIGMOD Conference on Management of Data, 1998.

12) S. Thomas, S. Sarawagi. Performance Evaluation and Optimization of Join Queries for Association Rule Mining. In Proc. of First International Conference on Data Warehousing and Knowledege Discovery (DAWAK), 1999.

13) T. Tamura, M. Oguchi, M. Kitsuregawa. Parallel Database Processing on a 100 Node PC Cluster: Cases for Decision Support Query Processing and Data Mining. In Proc. of SC97: High Performance Networking and Computing, 1997.

14

Secondary Storage Configurations for Advanced Data Engineering

Haruo Yokota
Ryota Abe
Tokyo Institute of Technology

ABSTRACT

Applications of advanced data engineering require scalable and reliable secondary storage systems. These requirements make the system configurations storage-centric rather than CPU-centric. To implement a scalable storage-centric system, network attached storage (NAS) and storage area network (SAN) architectures have recently attracted a great deal of attention. Current technological progress also allows disk-resident data processing. This capability is useful for managing distributed disks, as well as for executing application programs on the disks, such as the proposed intelligent disks e.g., active disks. We propose autonomous disks for self management in network environments using their disk-resident data processing capability. Autonomous disks can handle disk failures and load skews by a combination of active rules and distributed directories while remaining transparent to the hosts. These will be key functions for the next generation of storage systems, and are applicable to many advanced applications, such as a large Web server having many HTML files and B2B e-commerce frameworks generating enormous XML files. In this chapter, we describe the concept of autonomous disks by comparing them with ordinary disks.

1 Introduction

Improvements in computer architecture and Internet technology have enabled the implementation of many types of sophisticated advanced database applications. We already have high performance processors and high bandwidth networks, which are adequate to realize these. Compared with these components, however, the progress in performance enhancement of secondary storage systems is rather poor. Hence, the scalability and reliability of secondary storage systems have become amongst the most significant aspects for high performance intensive data processing.

It is true that recording densities of magnetic hard disk drives have recently improved tremendously. However, a large capacity hard disk drive has problems in performance and reliability. Access latency, caused by head seek and rotation wait, is an intrinsic restriction on the performance of disk devices. Each access for a large capacity disk induces latency, while parallel accesses for multiple small disks for the same amount of data could mask the latency. Simultaneously, the increase in capacity of a disk makes the effects of its failure more serious. From the above observations, it is preferable to use a large number of small disks to achieve storage I/O with adequate performance. These disks should also be shared by a number of hosts via some network.

From the point of view of system configuration the storage system becomes the center of the entire system. These are called *storage-centric* configurations . To implement a scalable storage-centric system, *network attached storages (NAS)* and *storage area network (SAN)* architectures have attracted a great deal of attention recently. A NAS is directly connected to an IP network to be shared by multiple hosts connected to a local area network (LAN), while a SAN consists of a dedicated network, separate from the LAN, using serial connection of storage devices, e.g., Fiber Channel.

In these configurations, disks are currently assumed to be passive devices. That is, all their behavior is controlled by their hosts via the network. Therefore, communications between disks and their hosts are very frequent and this limits the performance of the I/O. Moreover, to make the system efficient, placement of the data, including replicas, is very important. The management of data location requires a dedicated host. Because the host must also control all the accesses to the system the host will become a bottleneck as the system becomes large.

Furthermore, reliability of the total system largely depends on the reliability of the central host, the management software on it, and its maintenance operations. It is well known that the reliability of software and operations are rather low compared with hardware [1]. These reliabilities multiply the base reliability of disks, because the system is a series configuration for reliability calculation. This means the reliability of the system is less than that of the least reliable component. There are several approaches to preventing loss of data due to a disk failure, such as mirroring, using error correcting codes, and applying parity calculation techniques, as described in papers on RAID [2, 3]. However, these approaches are less flexible.

2 Autonomous Disks

To make a storage-centric system scalable, by removing the performance bottlenecks, and reliable, by excluding the complicated central control, distributed autonomous control of the storage nodes is essential.

Timely disk-resident data processing has recently attracted a great deal of attention. Technological progress, such as compact high-performance microprocessors for disk controllers and large semiconductor memories for disk cache, allow this capability. We propose a concept of *autonomous disks*, which utilize disk-resident processing capability to manage the data stored in them [4].

There are several other academic research projects to utilize disks' capability for executing application programs: the IDISK project at UC Berkeley [5] and the Active Disk projects at Carnegie Mellon [6] and UC Santa Barbara/Maryland [7]. They focus on the functions and mechanisms for making a combination of a disk-resident processor and a host execute storage-centric user applications, such as decision support systems, relational database operations, data mining, and image processing. However, they do not consider the management and reliability of the data.

A set of autonomous disks is configured as a cluster in a network for either a NAS or a SAN. Data is distributed within the cluster so it can be accessed uniformly. The disks accept simultaneous accesses from multiple hosts via the network. Disk-resident processors in the disks handle data distribution and load skews to enable efficient data processing. They are also capable of tolerating disk failures and software errors in the disk controllers, and reconfiguring the cluster after the damaged disks are repaired. The data distribution, skew handling, and fault tolerance are completely transparent to their hosts. Thus, the hosts are not involved in communication between disks to realize these functions. This provides high scalability of the system without the central controller.

2.1 Approaches to Implementation

There are many approaches to achieving the capabilities described above. We propose to use rules with command layers and a stream interface [4]. They provide sufficient flexibility to the user. Moreover, to provide transparency to the hosts, our approach adopts a distributed directory combined with the rules. It enables each disk to accept requests for all the data in a cluster and balances the load within the cluster.

2.1.1 Rule Description

The autonomous disks are controlled by active rules, which are also called Event-Condition-Action (ECA) rules [8, 9]. The syntax we use for a rule here is very simple and common in active databases (we can omit the `else` clause):

```
when  (an event triggering a rule);
if    (conditions for checking the state);
then  (actions executed if the conditions are satisfied);
else  (actions executed if the conditions are not satisfied).
```

By using a programming language, all the functions can be implemented within the disk controllers. The main reason for adopting rules is to provide user-friendly descriptions for controlling distributed disks and to restrict the capabilities allowed to users. If we allow users to write anything in an ordinary programming language, they may violate some limit that will destroy some important information for controlling the disk because of confusion as to how to describe what they wish to do. The rule descriptions permit only combinations of the basic commands provided, but are sufficiently flexible to manipulate streams and adapt the state of a disk to suit. Basic commands are implemented by a programming language, and combined together by the rules to realize the higher interface commands and autonomy properties of the disks.

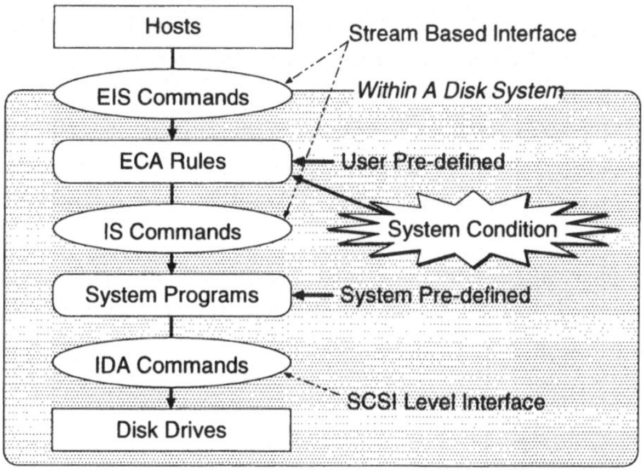

Figure 1: Command Layers

For example, there are several strategies for fault tolerance, such as how many faults should be tolerated, or how often the synchronization function is invoked between the replications. If the strategies are specified by rules, a user can easily change strategies simply by changing the conditions and the combination of the commands provided. The conditions are types of streams, states of streams, states of the system, and so on. Therefore, strategies can be specified independently for each stream.

2.1.2 Command Layers

We introduce three command layers into the autonomous disks: *external interface stream (EIS) commands*, *internal stream (IS) commands*, and *internal disk access (IDA) commands.* Figure 1 depicts the relationship between them. The IDA commands are the ordinal seek, read, and write commands for a disk, such as SCSI or ATA commands. The IS commands provide the stream interface for the rule descriptions, while the EIS commands provide the stream interface for the hosts.

The stream-based interface approach of the EIS and IS commands allows the treatment of not only simple disk blocks but also many types of interface units. For instance, the most popular interface unit in current operating systems is a file that can be simply treated as a stream. The National Storage Industry Consortium (NSIC) also proposes stream-based commands for the next generation of storage devices [10]. The flexible interface unit is essential for advanced data engineering.

The following three commands are basic EIS commands:

- `retrieve(HostID, StreamID)`
- `insert(HostID, StreamID, Stream)`
- `delete(HostID, StreamID)`

The `retrieve` command searches for a stream having the identifier `StreamID` within the cluster and returns the stream to the host indicated by `HostID`. If the stream cannot be found within the cluster, it returns false as a return value. The `insert` command inserts the `Stream` into a place corresponding to the `StreamID`. When the insertion succeeds, true is returned to the host `HostID`. The `delete` command searches for the stream of `StreamID` then deletes it, and returns the status to the host `HostIS`.

An EIS command is transformed into combinations of IS commands by pre-defined ECA rules together with the conditions of the system. The following are examples of IS commands:

- `send(Destination,Stream)`
- `traverse_directory(StreamID)`
- `insert_local(Location,StreamID,Stream)`
- `delete_local(Location,StreamID)`
- `mapping(DiskID,Type)`
- `compose(Substreams)`
- `decompose(Stream)`

The IS commands are not limited to these. If new functions are required, new IS commands can be introduced. Basically, these commands are pre-defined by the system in some programming language, including the IDA commands. Occasionally, an IS command may be implemented by a combination of other IS commands and the ECA rules.

To deal with heterogeneity, implementation of the IS commands can be varied for each disk. Note that the IS commands should not affect the basic disk control function of protecting the disk from illegal use of the commands.

The `send` IS command sends a stream to a destination. Because the disk identifier is known at the IS command level, the destination is either `HostID` or `DiskID`. There is no difference between disk-host and disk-disk communication at this level.

The commands `insert_local` and `delete_local` execute insert and delete operations on the stream at the required location in the disk. For fault tolerance, data should be replicated on several disks. Because a tightly synchronous update of backups has a large overhead, it is better to update the replicas asynchronously. Sequential log files can be used to implement the asynchronous update of backups. For these purposes, meta-information is required to map the disks, keeping primary, backup and logs. The command `mapping` returns a copy of the mapping information kept in each disk.

A stream can be fragmented into multiple sub-streams stored in different disks to enhance disk throughput. The command `decompose` fragments a stream into substreams, and `compose` regenerates the original stream from the fragmented sub-streams.

2.1.3 Distributed Directory

It is crucial for disk autonomy to make each disk accept EIS commands directly from any of the hosts. For this uniformity, each disk has part of a distributed directory. The **traverse_directory** IS command returns an identifier of the disk containing the stream and its location on the disk.

The Fat-Btree is proposed as a distributed directory structure capable of supporting a high-speed access method and a load-balancing mechanism among the disks [11]. In the Fat-Btree, the lower level index nodes, which have relatively high update frequencies, have few copies while the upper level index nodes, which have low update frequencies, are copied to many disks. Therefore, the Fat-Btree also has the advantage of lower synchronization overhead when the Btree structure is modified by repeated update operations.

2.2 Behavior of Autonomous Disks

2.2.1 Retrieving with Distributed Directory

In this subsection, let us consider the ECA rules to explain the behavior of autonomous disks. The following rule treats the **retrieve** EIS command:

```
Rule_1:
 when retrieve(HostID, StreamID);
 if   (D = traverse_directory(StreamID)).disk == Own
      and D.type == single;
 then send(HostID, D.stream).
```

When a disk receives a **retrieve** command via the network, it traverses its distributed directory, such as the Fat-Btree, to derive the location of the stream indicated by StreamID. If the result of the directory traversal indicates that the stream is stored within the disk, it returns the stream to the host identified as HostID. If the stream does not exist on the disk, it transmits the command to a disk, derived from the directory traversal, which may contain the stream. The following rule describes this operation, or can be combined with Rule_1 using an else clause.

```
Rule_2:
 when retrieve(HostID, StreamID);
 if   (D = traverse_directory(StreamID)).disk != Own;
 then send(D.disk, retrieve(HostID, StreamID)).
```

When the target stream is fragmented into a number of disks, the stream identified by StreamID is not the target stream itself, but is a list of identifiers for the sub-streams. The rule for searching a fragmented stream is defined as:

```
Rule_3:
 when retrieve(HostID, StreamID);
 if   (D = traverse_directory(StreamID)).disk== Own
```

```
      and D.type == fragmented;
 then send(HostID, compose(retrieve_list(Own, D.stream))).
```

The `retrieve_list` internal stream command searches for sub-streams according to the list `D.stream`. These sub-streams are finally combined as a stream and returned to the host.

These three rules, `Rule_1`, `Rule_2`, and `Rule_3`, are triggered at the same time, and check their conditions. Hereafter, for simplicity, we concentrate on the case where the stream is not fragmented. However, the treatment of the fragmented stream is easily installed by appending rules such as `Rule_3`.

2.2.2 Asynchronous Update of Backups

Rules for the `insert` and `delete` commands are defined similarly to those for the `retrieve` command. However, if the target stream of these update commands is replicated, to tolerate faults, the replicas must be updated as well. Backups of the stream must be placed on different disks to the primary, but each disk will contain both primaries and backups of different streams. Because the synchronization cost makes simultaneous update of these replicas expensive, asynchronous update of backups using sequential log files is more cost-effective. To implement this, the following rule is defined:

```
Rule_4:
 when insert(HostID, StreamID, Stream);
 if   (D = traverse_directory(StreamID)).disk == Own;
 then L = mapping(Own, log),
      lock(StreamID),
      send(L.disk, put_log(D, insert(HostID, StreamID, Stream))),
      insert_local(D.location, StreamID, Stream),
      unlock(StreamID),
      send(HostID, true);
 else send(D.disk, insert(HostID, StreamID, Stream)).
```

`Rule_4` first checks the location of the stream as does the rule of the `retrieve` command. If the stream is not stored in the current disk it transmits the command to an appropriate disk, derived from the directory traversal. If the disk is the correct one to store the stream, it derives the identifier of another disk for storing its log by using the `mapping` command, sends the log information to the log disk, inserts the stream into the appropriate location with the `insert_local` command, and returns the value true as the success flag. Backups are not updated at this point. If the volume of log information exceeds a threshold, the log information is transferred to backup disks and the contents of the backups are brought up to date. The procedure is implemented by the following rule. The disk identifier of the backup is also derived by the `mapping` command.

```
Rule_5:
```

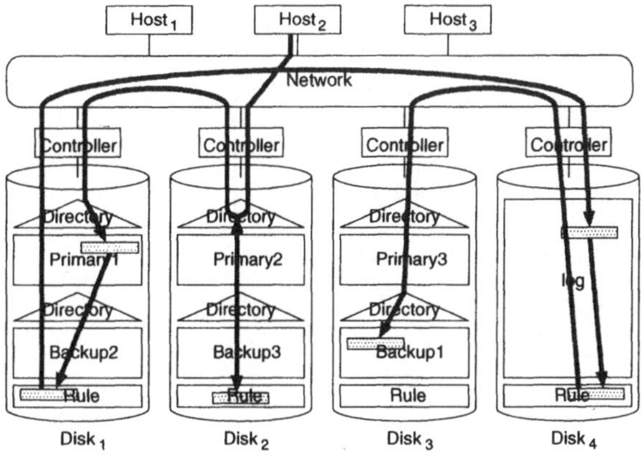

Figure 2: Data flow in autonomous disks

```
when  put_log(D, CMD);
if    count(log, D) > Threshold;
then  insert_local(log(D, CMD)),
      B = mapping(D, backup),
      catch_up(D, B);
else  insert_local(log(D, CMD)).
```

Figure 2 illustrates an example of data flow for an insert operation. In this example, Host_2 submits a request to insert a stream to Disk_2 via the network. However, the stream should be stored into Disk_1 for proper data distribution. **Rule_4** is triggered to derive the destination disk, Disk_1, using the **traverse_directory** command, and to transmit the request to Disk_1 by the **else** clause of **Rule_4**. In Disk_1, **Rule_4** is triggered again, but it executes an actual insert operation instead of transmitting the request, because **traverse_directory** indicates that the current disk is appropriate. During the insertion process, Disk_1 sends log information to the log disk (Disk_4) indicated by the mapping information according to the write ahead log (WAL) protocol, and executes the internal stream command **insert_local**. Finally it returns the value true to Host_2.

Update logs accumulate gradually in Disk_4, and each put-log operation triggers a check operation on the log size. When the log size exceeds a threshold, catch-up operations are asynchronously invoked by **Rule_5**. In these operations, the accumulated logs are transmitted and applied to the corresponding backup disk (Disk_3).

2.2.3 Handling Disk Failures or Errors

If there are disk failures or software errors that stop the disk controller, traversal of the distributed directory will be interrupted. In that case, the following rule will be

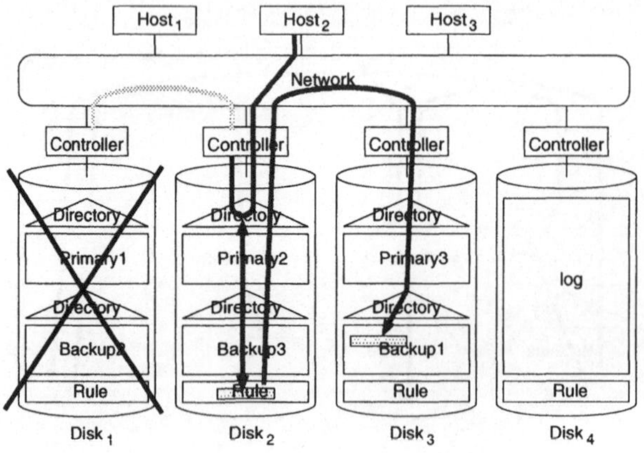

Figure 3: Fault treatment in autonomous disks

invoked.

```
Rule_6:
 when failure(retrieve(HostID, StreamID));
 if   B = mapping((D = traverse_directory(StreamID)), backup);
 then catch_up(D, B),
      send(B, retrieve_backup(HostID, StreamID)).
```

The rule finds a backup disk using mapping information, invokes a catch-up operation for the backup disk, and continues the traverse operation on the backup disk. The flow of the fault treatment is illustrated in Figure 3. In this example, $Disk_2$ tries to transmit a retrieve command to $Disk_1$, but $Disk_1$ cannot be accessed because of a disk failure, or software errors, in the disk's controller. Then `Rule_6` invokes a catch-up operation on the backup disk ($Disk_3$), and transmits the retrieve command to it.

2.2.4 Handling Skews

Distribution skew is basically handled by the distributed directory. The Fat-Btree is capable of handling distribution skew. It detects the skew, and migrates overloaded data to the logical neighboring disks. Therefore, the skew is systolically dissipated through the cluster. Algorithms for detecting skew and migrating data are presented in [11]. Cluster reconfiguration after replacing a failed disk with a new disk is a special case of skew handling for the Fat-Btree. The new empty disk is gradually filled with migrated streams.

2.3 Varying Strategies

In this example, the catch-up operation is invoked only when the size of the pre-applied log exceeds a threshold. We can modify the strategy by changing the rules. For example, we can use load skew information to trigger the catch-up operation, or make the backup and log disks wait until their workload is low.

We can also vary the strategies for fault tolerance, but only by changing the rule definitions. The log can also be duplicated to improve the reliability of the system. `Rule_4` can be changed to treat double logs as follows:

```
Rule_4':
 when insert(HostID, StreamID, Stream);
 if   (D = traverse_directory(StreamID)).disk == Own
      and  D.type == double_log;
 then L1 = mapping(Own, log1),
      L2 = mapping(Own, log2),
      lock(StreamID),
      send(L1.disk, put_log(D, insert(HostID, StreamID, Stream))),
      send(L2.disk, put_log(D, insert(HostID, StreamID, Stream))),
      insert_local(D.location, StreamID, Stream),
      unlock(StreamID),
      send(HostID, true);
 else send(D.disk, insert(HostID, StreamID, Stream)).
```

In this example, the update logs are sent and written to the two disks indicated by the mapping information. We can also easily change the number of backup disks:

```
Rule_5':
 when put_log(D, CMD);
 if   count(log, D) > Threshold
      and  D.type == double_backup;
 then insert_local(log(D, CMD)),
      B1 = mapping(D, backup1),
      catch_up(D, B1),
      B2 = mapping(D, backup2),
      catch_up(D, B2);
 else insert_local(log(D, CMD)).
```

The increase of replication is effective for concurrent retrieval operations. The replication of a stream is transparent to the hosts, except for the administrator host. The replication information is stored as a property of the stream in the distributed directory.

Fragmentation of a stream is also transparent to the hosts. The size of the stream, or the access patterns for the stream, are used as criteria for fragmentation. The decomposition operation of the stream is also treated by rules. These strategies are varied according to the circumstances of the stream. In other words, each stream

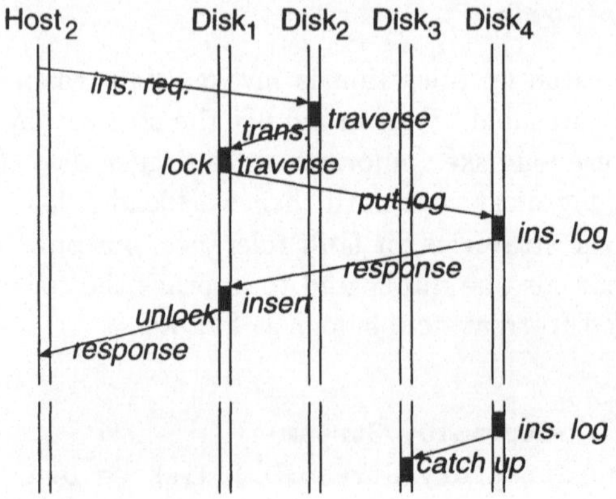

Figure 4: Communication Structure using Autonomous Disks

can have its own strategy. This is a desirable benefit derived by adopting the rule mechanism.

3 Comparison with Ordinary Disks

3.1 Communication Structure

The communication structure of the example in the previous section is depicted in Figure 4. It is easy to see that the whole process is independent of the host except for the first command submission and the receipt of the response. In contrast, the host must control every disk I/O if we want to execute the same function using ordinary disks. The communication structure of the same situation using ordinary disks is depicted in Figure 5. These two figures also demonstrate that total communication on the network is reduced with the autonomous disks.

If we use autonomous disks, the target stream is locked and unlocked in the disk storing the stream for concurrency control. Any hosts desiring to access the stream will reach the disk via the distributed directory, and will try to acquire a lock on the stream in the disk. Thus, no extra communication for controlling concurrency is required for autonomous disks. On the other hand, each host must manage locks to access the clustered disks as in Figure 5. If a host wants to access a disk page in the cluster it communicates to all the other hosts, or a host associated with the disk page, to try to acquire a lock on the page.

3.2 Cost Estimation

We estimate costs for basic operations of ordinary and autonomous disks. For the estimation, the following parameters are introduced.

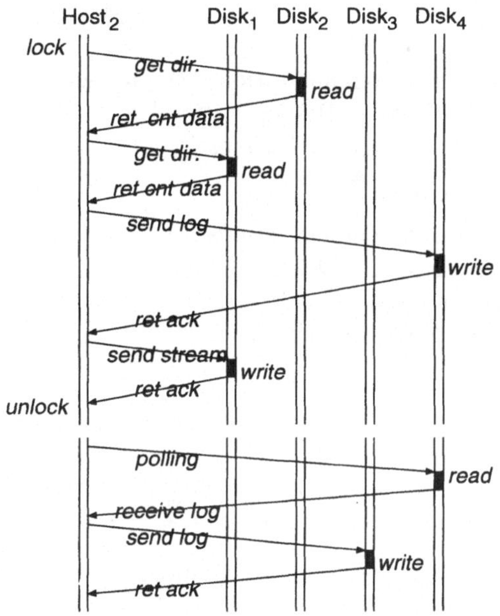

Figure 5: Communication Structure using Ordinary Disks

- D_c: Data size of control information
- D_s: Data size of a target stream
- N_d: The number of disks, excluding log disks
- N_s: The number of stored streams
- H: Height of a Fat-btree stored into disks
- $T_{rw}(D)$: Time for read or write in a disk
- $T_{cm}(D)$: Time for communication between a host and a disk

The computing power also affects system performance, but it is negligible compared with costs of disk access and network communication. Therefore, we ignore the cost of computation in this estimation. We also assume that the structure of the Fat-Btree is not changed by insert operations in this estimation, and that backup disks can be decided uniquely.

3.2.1 The Cost of Operations in Ordinary Disks

Insert: The followings are the main costs for an insert operation using ordinary disks.

- Cost of searching for the destination disk to insert the stream
  - Count of host-disk communications (D_c) for traversing index: $2H$

  - Count of disk accesses (D_c) for traversing index: H
- Cost of taking logs
  - Count of host-disk communications (D_s) for sending log: 1
  - Count of disk accesses (D_s) for writing log: 1
  - Count of host-disk communications (D_c) for sending ack: 1
- Cost of inserting a stream
  - Count of host-disk communications (D_s) for sending stream: 1
  - Count of disk accesses (D_s) for writing log: 1
  - Count of disk accesses (D_c) for updating directory: 1
  - Count of host-disk communications (D_c) for sending ack: 1
- Cost of backup
  - Count of host-disk communications for invoke and ack backup (D_c): 2
  - Count of disk accesses (D_s) for reading and writing log: 2
  - Count of disk accesses (D_c) for reading and writing directory: 2
  - Count of host-disk communications (D_s) for sending log: 2

From the above, we can derive a cost formula for the total cost of an insert operation with synchronous backup:

$$\begin{aligned} T_{insert_{os}} &= 2H \times T_{cm}(D_c) + H \times T_{rw}(D_c) && \text{(traverse index)} \\ &\quad + T_{cm}(D_s) + T_{rw}(D_s) + T_{cm}(D_c) && \text{(insert log)} \\ &\quad + T_{cm}(D_s) + T_{rw}(D_s) + T_{rw}(D_c) + T_{cm}(D_c) && \text{(insert stream)} \\ &\quad + 2T_{cm}(D_c) + 2T_{rw}(D_s) + 2T_{rw}(D_c) + 2T_{cm}(D_s) && \text{(backup)} \\ &= (2H+4)T_{cm}(D_c) + 4T_{cm}(D_s) + (H+3)T_{rw}(D_c) + 4T_{rw}(D_s) && (14.1) \end{aligned}$$

On the assumption of sequential execution for the whole process, the cost becomes the response time of an insert operation. On the other hand, if we adopt asynchronous backup, we can omit the time for backup from the response time of insertion.

$$T_{insert_{oa}} = (2H+2)T_{cm}(D_c) + 2T_{cm}(D_s) + (H+1)T_{rw}(D_c) + 2T_{rw}(D_s) \quad (14.2)$$

Delete: The cost formula for a delete operation is similar to that for an insert. The treatment of log and directory are different. For a delete operation with synchronous backup:

$$\begin{aligned} T_{delete_{os}} &= 2H \times T_{cm}(D_c) + H \times T_{rw}(D_c) && \text{(traverse index)} \\ &\quad + T_{cm}(D_c) + T_{rw}(D_s) + T_{cm}(D_s) && \text{(get before-image)} \\ &\quad + T_{cm}(D_s) + T_{rw}(D_s) + T_{cm}(D_c) && \text{(insert log)} \\ &\quad + T_{cm}(D_c) + T_{rw}(D_c) && \text{(delete stream)} \\ &\quad + 4T_{cm}(D_c) + 2T_{rw}(D_c) && \text{(backup)} \\ &= (2H+7)T_{cm}(D_c) + 2T_{cm}(D_s) + (H+3)T_{rw}(D_c) + 2T_{rw}(D_s) && (14.3) \end{aligned}$$

Table 1: The number of nodes in a Btree for various numbers of streams

N_s	H	$N_n(1)$	$N_n(2)$	$N_n(3)$	$N_n(4)$	$N_n(5)$
100K	3	1	6	722		
1M	3	1	53	7214		
10M	4	1	4	521	72135	
100M	4	1	38	5204	721348	
1G	5	1	3	376	52035	7213476

Similarly to the insert operation, the response time of a delete operation can be reduced by asynchronous backup:

$$T_{delete_{oa}} = (2H+3)T_{cm}(D_c) + 2T_{cm}(D_s) + (H+1)T_{rw}(D_c) + 2T_{rw}(D_s) \quad (14.4)$$

Retrieve: Because a retrieve operation does not require logs and backups, the cost of taking log and backup should be eliminated. Therefore, the cost formula becomes:

$$\begin{aligned} T_{retrieve_o} &= 2H \times T_{cm}(D_c) + H \times T_{rw}(D_c) & \text{(traverse index)} \\ &\quad + T_{rw}(D_s) + T_{cm}(D_s) & \text{(retrieve stream)} \\ &= 2H \times T_{cm}(D_c) + T_{cm}(D_s) + H \times T_{rw}(D_c) + T_{rw}(D_s) & (14.5) \end{aligned}$$

Of course, by assuming some cache mechanism, those costs will be reduced further. For instance, disk accesses during index traversal can be omitted by using a cache. In that case the response times for insert, delete and retrieval operations become as follows:

$$\begin{aligned} T_{insert_{oc}} &= (2H+2)T_{cm}(D_c) + 2T_{cm}(D_s) + T_{rw}(D_c) + 2T_{rw}(D_s) & (14.6) \\ T_{delete_{oc}} &= (2H+3)T_{cm}(D_c) + 2T_{cm}(D_s) + T_{rw}(D_c) + 2T_{rw}(D_s) & (14.7) \\ T_{retrieve_{oc}} &= 2H \times T_{cm}(D_c) + T_{cm}(D_s) + T_{rw}(D_s) & (14.8) \end{aligned}$$

3.2.2 The Cost of Operations in Autonomous Disks

First, we derive the probability that a disk contains the target node, to estimate the frequency of communication between disks. If we adopt a Fat-btree as the distributed directory, the probability that a PE has a node in level h that is necessary for the traverse is:

$$p(h) = \frac{N_n(h) + N - 1}{N \times N_n(h)}$$

Here, $N_n(h)$ is the total number of nodes for level h in a Btree. Table 1 lists values for a variety of stream numbers N_s.[11)]

Insert: At this time, the cost of an insert operation:

- Count of host-controller communications (D_s) for sending stream: 1
- Cost of searching for the destination disk to insert the data
  - Count of controller-controller communications (D_c) for traversing index: $\sum_{h=1}^{N-1}(1-p(h))$
  - Count of disk accesses (D_c) for traversing index: H
- Cost of taking logs
  - Count of controller-controller communications (D_s) for sending log: 1
  - Count of disk accesses (D_s) for writing log: 1
  - Count of controller-controller communications (D_c) for sending ack: 1
- Cost of inserting a stream
  - Count of disk accesses (D_s) for writing stream: 1
  - Count of disk accesses (D_c) for updating directory: 1
  - Count of host-controller communications (D_c) for sending ack: 1
- Cost of backup
  - Count of controller-controller communications (D_s) for sending log: 1
  - Count of disk accesses (D_s) for reading and writing log: 2
  - Count of disk accesses (D_c) for reading and writing directory: 2

From the above, we can derive cost formulae for the response times for an insert operation with synchronous backup, asynchronous backup, and index-cached asynchronous backup:

$$\begin{aligned}
T_{insert_{as}} &= T_{cm}(D_s) && \text{(send stream)} \\
&+\left\{\sum_{h=1}^{N-1}(1-p(h))\right\} T_{cm}(D_c) + H \times T_{rw}(D_c) && \text{(traverse directory)} \\
&+T_{cm}(D_s) + T_{rw}(D_s) + T_{cm}(D_c) && \text{(insert log)} \\
&+T_{rw}(D_s) + T_{rw}(D_c) + T_{cm}(D_c) && \text{(insert stream)} \\
&+T_{cm}(D_s) + 2T_{rw}(D_s) + 2T_{rw}(D_c) && \text{(backup)} \\
&= \left\{\sum_{h=1}^{N-1}(1-p(h)) + 2\right\} T_{cm}(D_c) + 3T_{cm}(D_s) + (H+3)T_{rw}(D_c) + 4T_{rw}(D_s)
\end{aligned} \tag{14.9}$$

$$T_{insert_{aa}} = \left\{\sum_{h=1}^{N-1}(1-p(h)) + 2\right\} T_{cm}(D_c) + 2T_{cm}(D_s) + (H+1)T_{rw}(D_c) + 2T_{rw}(D_s) \tag{14.10}$$

$$T_{insert_{ac}} = \left\{\sum_{h=1}^{N-1}(1-p(h)) + 2\right\} T_{cm}(D_c) + 2T_{cm}(D_s) + T_{rw}(D_c) + 2T_{rw}(D_s) \tag{14.11}$$

Delete: As with ordinary disks, the cost of a delete operation is similar to that of an insert operation. The costs of a delete operation with synchronous backup, asynchronous backup, and index-cached asynchronous backup are:

$$
\begin{aligned}
T_{delete_{as}} &= T_{cm}(D_c) && \text{(invoke delete)} \\
&+ \left\{ \sum_{h=1}^{N-1} (1 - p(h)) \right\} T_{cm}(D_c) + H \times T_{rw}(D_c) && \text{(traverse directory)} \\
&+ 2T_{rw}(D_s) + T_{cm}(D_s) + T_{cm}(D_c) && \text{(insert log)} \\
&+ T_{rw}(D_c) + T_{cm}(D_c) && \text{(delete stream)} \\
&+ T_{cm}(D_c) + 2T_{rw}(D_c) && \text{(backup)} \\
&= \left\{ \sum_{h=1}^{N-1} (1 - p(h)) + 4 \right\} T_{cm}(D_c) + T_{cm}(D_s) + (H + 3)T_{rw}(D_c) + 2T_{rw}(D_s)
\end{aligned} \quad (14.12)
$$

$$
T_{delete_{aa}} = \left\{ \sum_{h=1}^{N-1} (1 - p(h)) + 3 \right\} T_{cm}(D_c) + T_{cm}(D_s) + (H + 1)T_{rw}(D_c) + 2T_{rw}(D_s) \quad (14.13)
$$

$$
T_{delete_{ac}} = \left\{ \sum_{h=1}^{N-1} (1 - p(h)) + 3 \right\} T_{cm}(D_c) + T_{cm}(D_s) + (H + 1)T_{rw}(D_c) + 2T_{rw}(D_s) \quad (14.14)
$$

Retrieve: The following are the response times for a retrieve operation without and with index caching:

$$
T_{retrieve_a} = \left\{ \sum_{h=1}^{N-1} (1 - p(h)) + 3 \right\} T_{cm}(D_c) + 2T_{cm}(D_s) + H \times T_{rw}(D_c) + T_{rw}(D_s) \quad (14.15)
$$

$$
T_{retrieve_{ac}} = \left\{ \sum_{h=1}^{N-1} (1 - p(h)) + 3 \right\} T_{cm}(D_c) + 2T_{cm}(D_s) + T_{rw}(D_s) \quad (14.16)
$$

3.2.3 Comparisons of Estimated Costs

We now compare the cost of an insert, a delete and a retrieve operation using autonomous disks with those using ordinary disks. For a given operation, for instance an insert operation, the costs of disk accesses are equivalent between autonomous disks and ordinary disks under the same condition: $T_{insert_{os}}$ vs. $T_{insert_{as}}$, $T_{insert_{oa}}$ vs. $T_{insert_{aa}}$, and $T_{insert_{oc}}$ vs. $T_{insert_{ac}}$. Thus, only the costs related to communication are different. The same argument applies to delete and retrieve operations. By considering their operations, this is easily understood.

We now estimate response times for those operations by substituting some practical values for the variables in the previous cost formulas. Using these substitutions,

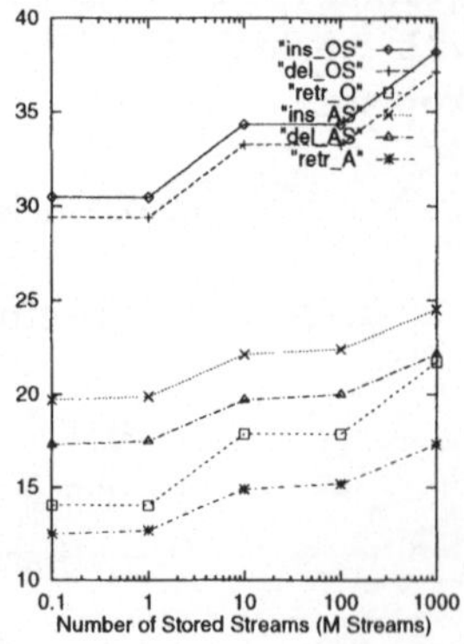

Figure 6: Response time with synchronous backup

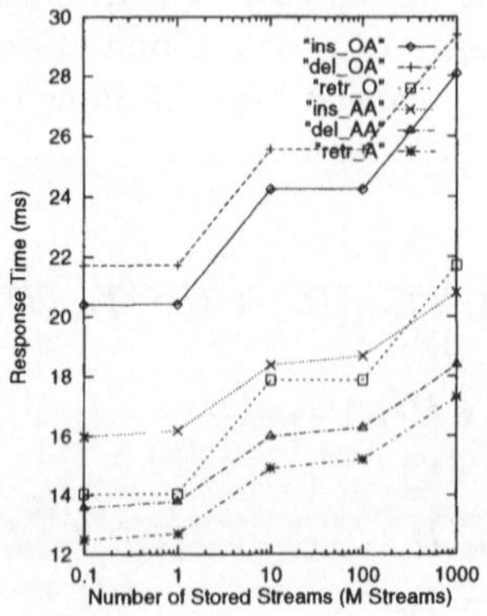

Figure 7: Response time with asynchronous backup

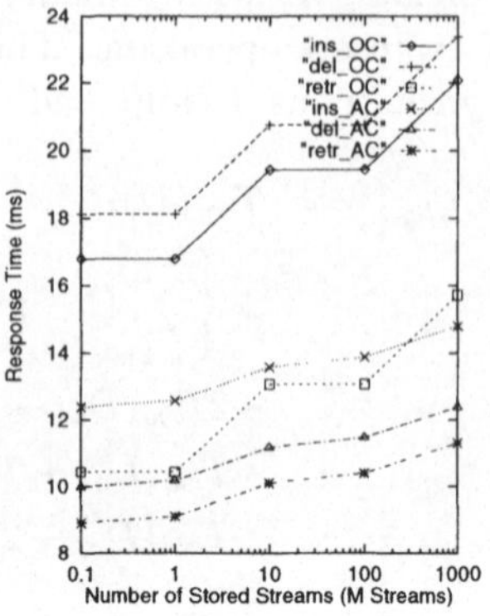

Figure 8: Response time with cached async. bkup

disk access time and communication time should be clarified. The disk access time $T_{rw}(D)$ is composed of average disk access latency $T_{accessletency}$ and data transfer time D/BW_{disk}, where BW_{disk} denotes the disk bandwidth. The communication time $T_{cm}(D)$ is also composed of the average network setup time T_{setup} and data transfer time D/BW_{net}, where BW_{net} denotes the network bandwidth.

$$T_{rw}(D) = T_{accessletency} + D/BW_{disk} \quad (14.17)$$

$$T_{cm}(D) = T_{setup} + D/BW_{net} \quad (14.18)$$

As an example, we substitute the following values: $BW_{net} = 100\text{Mbps}$, $T_{setup} = 1\text{ms}$, $BW_{disk} = 20\text{MB/s}$, $T_{accessletency} = 1\text{ms}$, $D_c = 4\text{KB}$, $D_s = 4\text{KB}$, $N_d = 8$, and $N_s = 512$. Figures 6 to 8 compare the response times of insertion, deletion and retrieval operations, respectively. Those graphs indicate that the autonomous disks significantly improve access performance of each operation as well as providing rich functionality.

These three graphs also show that, in every situation, the times for retrieve operations and update operations using autonomous disks are less than those for ordinary disks. This demonstrates that the reduction of communication costs is particularly effective in update operations because it affects communication for the log and backup.

4 Concluding Remarks

We have proposed a concept of an autonomous disk that is an enhancement of a disk connected to a storage area network (SAN) or a network-attached-storage (NAS) drive, having high transparency, flexibility, and scalability. Those features must be the key to the next generation secondary storage systems.

Data is distributed within a cluster of autonomous disks, and can be uniformly accessed from multiple hosts simultaneously. The disks are also able to handle data distribution, load skews and disk failures or errors in the disk controllers autonomously, and can reconfigure clusters after a repair. These aspects are accomplished within the cluster of autonomous disks without interaction with the hosts. This provides excellent transparency to the hosts.

To implement autonomous disks, we utilize a distributed directory effectively, and define three command layers: external interface stream (EIS) commands, internal stream (IS) commands, and internal disk access (IDA) commands. The EIS commands are implemented using the IS commands and ECA rules. The combination of command layers and rules enables users to describe the behavior of the autonomous disks easily. Moreover, the combination also provides flexibility in varying management strategies for individual streams. Thus, hosts are not involved in communication between disks to realize these functions. This provides high scalability of the system.

The transparency, flexibility, and scalability derived from this approach are essential. The variety of functions provided for hosts and disks allows the applications on the hosts to be independent of the disk configuration. Autonomous disks are effective in a scalable network environment containing a number of hosts and disks.

By estimating operation costs in autonomous and ordinary disks, we find that the autonomous disks reduce data transfer overheads compared to the same operations on ordinary disks. This indicates that the autonomous disks are suited to the network environment in respect to both functionality and performance.

Because we treat a file as a stream in the autonomous disks they are especially applicable to many advanced applications. For instance, Internet providers or web directory servers having a large number of HTML files have difficulty in maintaining files. Business-to-business e-commerce frameworks also generate enormous numbers of XML files. The proposed autonomous disks are capable of handling file streams very effectively for such situations, using the distributed directory. We can also provide a dedicated `search` EIS command to search such files for required components.

These functions can be implemented in PC clusters, although we intend to install them eventually in functional disks having disk-resident processors. We are currently implementing an experimental system using PCs connected by a network to evaluate the mechanism.

Bibliography

1) Jim Gray and Andreas Reuter. *Transaction Processing:Concepts and Techniques.* Morgan Kauf-Mann, 1993.

2) David A Patterson, Garth Gibson, and Randy H. Katz. A Case for Redundat Arrays of Inexpensive Disks(RAID). In *Proc. of ACM SIGMOD Conference*, pages 109–116, Jun 1988.

3) Peter M. Chen et al. RAID: High-Performance, Reliable Secondary Storage. *ACM Computing Surveys*, 26(2):145 - 185, Jun 1994.

4) Haruo Yokota. Autonomous Disks for Advanced Database Applications. In *Proc. of International Symposium on Database Applications in Non-Traditional Environments (DANTE'99)*, pages 441–448, Nov. 1999.

5) Kimberly Keeton, David A. Patterson, and Joseph M. Hellerstein. A Case for Intelligent Disks (IDISKs). *SIGMOD Record*, 27(3):42–52, Sep. 1998.

6) Erik Riedel, Garth Gibson, and Christos Faloutsos. Active Storage for Large-Scale Data Mining and Multimedia. In *Proc. of the 24th VLDB Conf.*, pages 62–73, 1998.

7) Anurag Acharya, Mustafa Uysal, and Joel Saltz. Active Disks: Programming Model, Algorithms and Evaluation. In *Proc. of the 8th ASPLOS Conf.*, Oct. 1998.

8) Dennis R. McCarthy and Umeshwar Dayal. The Architecture of an Active Data Base Management System. In *Proc. of SIGMOD Conf. '89*, pages 215–224, 1989.

9) J. Widom and S. Ceri (ed.). *Active Database Systems: Triggers and Rules for Advanced Database Processing.* Morgan Kaufmann Pub, 1996.

10) National Storage Industry Consortium (NSIC). Object based storage devices: A command set proposal. http://www.nsic.org/nasd/1999-nov/final.pdf, Nov 1999.

11) Haruo Yokota, Yasuhiko Kanemasa, and Jun Miyazaki. Fat-Btree: An Update-Conscious Parallel Directory Structure. In *Proc. of the 15th Int'l Conf. on Data Engineering*, pages 448–457, 1999.

15

Issues on Parallel Processing of Object Databases

Akifumi Makinouchi

Kyushu University

Tatsuo Tsuji

Fukui University

Hirofumi Amano

Kyushu University

Kunihiko Kaneko

Kyushu University

ABSTRACT

Traditional relational database systems are not adequate for storing, retrieving, and processing the data found in non-traditional database applications. Such data includes data for visualization, spatial data, and multimedia data, which is often used in World Wide Web pages. A number of applications requiring such structurally complex data are not only data-intensive but also cpu-intensive. Object Database Management Systems (ODBMS) are efficient to handle such data due to an abundance of data structures, rich functionality for data processing, and the ability to integrate programming language and databases. Issues involved in distributed and parallel processing of object queries and parallel object programming language are presented. An implementation of ODBMS based on ODMG3.0 standard is introduced also. First, parallel processing of path expressions is discussed. Path expressions are often used in ODB applications. Second, parallel processing using a proposed language is discussed, and methods of performance evaluation are presented. Finally, an ODBMS based on the ODMG3.0 standard with distributed and parallel processing ability is introduced. The system, named ShusseUo, consists of three components: WAKASHI, for distributed data storage; INADA, for ODMG3.0-compliant object handling; and WARASA, for distributed and parallel OQL compiler. The system architecture and functionality as well as a performance evaluation are presented.

1 Introduction

Traditional relational database systems are not adequate for storing, retrieving, and processing the data found for non-traditional database applications. Such data includes data for visualization, spatial data, and multimedia data, which is often used in World Wide Web pages. A number of applications requiring such structurally complex data are not only data-intensive but also cpu-intensive. Object Database Management Systems (ODBMS) are efficient to handle such data due to an abundance of data structures, rich functionality for data processing, and the ability to integrate programming languages and databases.

In this chapter, we focus on issues involved in distributed and parallel processing of object queries and parallel object programming languages. An implementation of ODBMS based on ODMG3.0 standard is also introduced. This chapter consists of four sections including this one.

In section 2, parallel processing of path expressions is discussed. Path expressions are often used in ODB applications. Fast processing of path expressions are crucial in fast retrieval of objects. Parallel execution of a path expression utilizing index in an asynchronous parallel computing environment is addressed. This section is written by Tatsuo Tsuji.

In section 3, a parallel object programming language is introduced. The above-mentioned non-traditional applications are data-intensive and cpu-intensive. The object programming language is useful in integrating the database retrieval and the data processing uniformly. Since such applications are cpu-intensive, improvement of the performance of these applications can be expected by parallel processing. Parallel processing using the proposed language is discussed, and methods of performance evaluation are presented. This section is written by Hirofumi Amano.

Finally, in section 4, an ODBMS based on the ODMG3.0 standard with distributed and parallel processing ability is introduced. The system, named ShusseUo, consists of three components: WAKASHI, for distributed data storage, INADA, for ODMG2.0-compliant object handling, and WARASA, for distributed and parallel OQL compiler. This section focuses on the system architecture and functionality. In addiction, a performance evaluation using a large spatial database is presented. This section is written by Kunihiko Kaneko.

2 Termination detection of parallel index retrieval for complex objects

A fast retrieval method for handling complex objects and corresponding effective implementation schemes are essential in order to keep up with the rapid growth of database applications in which large and complicated objects are handled. Employing parallelism is one promising approach for the fast retrieval of complex objects[3)4)5)]. Another effective approach is providing fast indexing techniques for a set of complex objects, including both aggregation and inheritance hierarchies[2)]. A third approach is a combination of the above two approaches.

By dividing a large index into sub-indexes and placing each sub-index on a separate machine, we can improve the efficiency of index operations through parallelism. In [1], a parallel retrieval algorithm was proposed for an index structure mapped on the hyper-cube parallel machine architecture. In [6], an optimizing method using the horizontal and vertical splitting index was proposed.

In these studies, each processor transfers its own retrieval results asynchronously and independently of the other processors. No communications among processors are required for controlling transfer. In such a case, detection of the overall termination of the parallel retrieval is important.

In the following, we propose termination detection schemes of extremely low time and space cost, based on the notion of *branch history*. In our schemes, the multi-indexing scheme is used, which allows us to take advantage of pipeline parallelism.

Index Splitting and Parallel Retrieval

Let a *path expression* of a complex object be denoted as $P = C_1 A_1 \ldots A_n$. Here, A_1 is an attribute of the class C_1. $A_j (j > 1)$ is an attribute of the class C_j, and the domain of the attribute A_{j-1} of C_{j-1} is the class C_j. Let the value of P be $o_1 o_2 \ldots o_{n+1}$. Here o_1 is an instance of C_1, and $o_j (j > 1)$ is the value of the attribute A_{j-1} of object o_{j-1}. o_j is assumed to be not a set. o_{n+1} is an object identifier or a simple value such as an integer or a string.

Let the set of values of P be denoted as O_p. When the value of the attribute A_j in the object o_j is o_{j+1}, the pair $< o_j, o_{j+1} >$ is called an *index element* of P. Here o_{j+1} is a *key value* and o_j is the corresponding *data value*. The index I_p of the path expression P is the set of all index elements specified as:

$$I_p = \{< o_j, o_{j+1} > | o_1 o_2 \ldots o_n \in O_p, 1 \leq j \leq n\}$$

The *extension* of an object o is defined as the set of index elements that includes objects referring to o directly or indirectly. Consider the following retrieval query:

> "Retrieve the set of objects in the class C_1 for which the value of the attribute A_n is s_n"

The *multi-indexing scheme* traverses the extension of s_n using I_p in the order $< o_n, s_n >, < o_{n-1}, o_n > \ldots < o_1, o_2 >$, and finally returns the set of o_1 as the retrieval result.

Index splitting

Let I_p be divided into the sub-indexes $\{V_1, \ldots, V_n\}$, where $V_j = \{< o_j, o_{j+1} > | o_1 o_2 \ldots o_{n+1} \in= O_p\} (1 \leq j \leq n)$. The set of sub-indexes $\{V_1, V_2, \ldots, V_n\}$ is called a *V-partition* of I_p. Let each V_j be further divided into m_j sub-indexes horizontally, each of which is denoted as $HV_i^j (1 \leq i \leq m_j)$, and maintained in the processor element PE_i^j (See Fig.1).

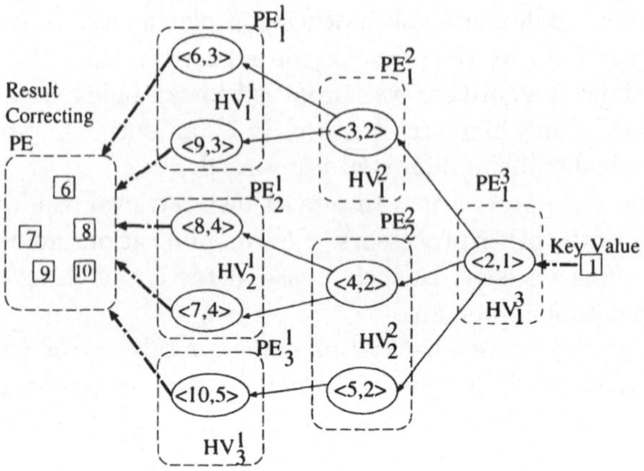

Figure 1: Index partition and parallel retrieval

Each PE acts as follows. If the PE receives a key value from an upstream neighbor PE (i.e., right-hand side PE in Fig.1), it searches the corresponding data values (OID) using the sub-index HV_i^j that it maintains. For each object in the result set of OIDs thus obtained, the PE sends the OID as a key value to one of the downstream neighbor PEs that maintains the related sub-index. If the PE maintains a portion of V_1(i. e. , HV_i^1), the set of OIDs searched in the PE is a portion of the required result OIDs. The PE transfers the results to the *result collecting PE*, which will be abbreviated as RCPE in the following.

Note that the above retrieval action is performed in parallel on each PE.

Termination Detection

When a retrieval request is issued, the retrieval will be performed in parallel along the horizontal paths of PEs. Such parallel retrieval in each PE does not require any synchronization mechanism with other PEs. Therefore, even if the resultant OIDs are reported from one of the PEs that stores a portion of V_1, these OIDs are only a portion of the final result, and other portions may be reported from other PEs subsequently. We will now describe the two termination detection schemes for such parallel index retrieval.

- **Object-level detection**

A retrieval *termination detection* scheme is outlined roughly for the simple example illustrated in Fig. 2. This scheme is simple enough that the basic concept should be easily understood. First, we describe the construction of the *branch history* attached to an object.

(i) Let the branch history attached to the object in the retrieval condition be "ϵ" (null string).

(ii) Let $< b_{i-1}, \cdots, b_1 >$ be the branch history of an object o transferred from the upstream neighbor PE, and let $R = \{o_1, o_2, \cdots, o_k\}$ be the retrieval result in the PE with o as the key value. Each OID in R is transferred to one of the downstream neighbor PEs with the attached branch history $< k, b_{i-1}, \cdots, b_1 >$.

The termination detection scheme uses a *table* T of branch histories to detect the acceptance of all the final retrieval results. T is held in RCPE (see Fig. 1) and all termination detection activities are performed in RCPE. Consider the example illustrated in Fig. 2, where the resultant OIDs and their *branch histories* are reported in the order $a \to b \to c \to d \to e$. Changes to T are shown in the right-hand side of Fig. 2.

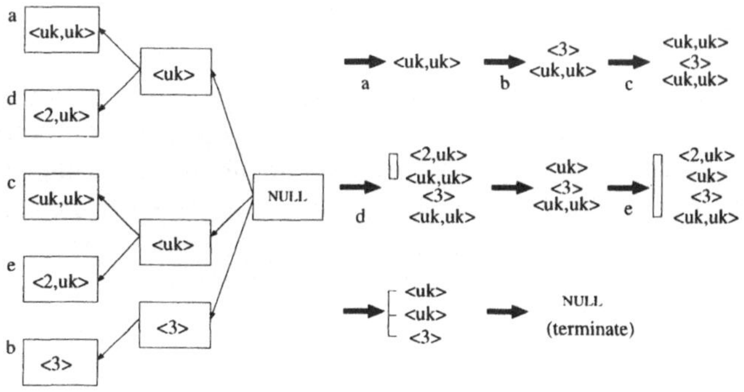

Figure 2: Object-level retrieval termination detection

For example, if the object a is received by RCPE, a's history, namely $< 2, 3 >$, is entered in T. When the object b is received, the associated history, $< 1, 3 >$, is merged alone into $< 3 >$, and is entered in T. When $< 2, 3 >$ is accepted by receiving c, a $< 2, 3 >$ can be found by searching T, and these two $< 2, 3 >$s are merged into a single $< 3 >$. Thus, T holds two $< 3 >$s. Next, the two $< 2, 3 >$s are accepted by receiving d and e they are merged into a single $< 3 >$. At this point, three $< 3 >$s remain in T, and these are merged into ϵ. Eventually, T becomes empty, and all portions of the final retrieval result can be concluded to have arrived at RCPE.

In this example, although the object that can be *logically* merged with a is d and not c, a can be merged with c, which arrives earlier than d, as is shown in Fig. 2. Moreover, d can be merged later with e, which has the same branch history. In addition however, in this case, the object that can be logically merged with d is a and not e. We can observe that since a, b, c, and e have the same branch history, two arbitrary histories can be merged according to the arrival order, which is independent of the logical references among the histories.

Note that the above-sketched scheme works in parallel with the group of PEs concerned in the retrieval.

- **Packet level detection**

Packet-level detection suffers from the following defects.

(1) Transferring overhead
Since a branch history is associated with an object (i.e., OID), the extra overhead of transferring the branch history through the communication line increases to a large degree. If the complex objects are deeply nested, this overhead would become very large.

(2) Termination detection overhead
While the index retrieval is performed in parallel by more than one PE, the termination detection is performed only in RCPE (See Fig.1). For every arrival of an object, the RCPE has to check the possibility of merging branch histories. This would be a great burden for the PE.

In order to reduce the above-described overhead, instead of attaching a branch history to every object, we attach the branch to every packet that stores more than one object for transfer. If a PE receives a packet P from one of the upstream neighbor PEs, for every object in P it searches the corresponding data values and packs them into an output packet, which is transferred to the related downstream neighbor PE. First, we modify the definition of branch history.

(i) Let the branch history attached to the packet including the object in the retrieval condition be "ϵ".

(ii) Let $< b_{i-1}, \cdots, b_1 >$ be the branch history of a packet P transferred from the upstream neighbor PE. The output packets would be produced one after another to store the retrieval results. Since the number 'p' packets produced is not known until all of the retrieval results are known, all packets except the final packet have the branch history $< unknown, b_{i-1}, \cdots, b_1 >$. Only when the last packet is fixed, p is known. Therefore, only the last packet has the branch history $< p, b_{i-1}, \cdots, b_1 >$.

For example, assume that the input packet P with the branch history $< 3, unknown >$ is received and that three packets are produced by retrieving each OID in P as a key value. The first two packets have the branch history $< unknown, 3, unknown >$ and the last packet has the branch history $< 3, 3, unknown >$. An example of the branch histories of packets is shown on the left-hand side of Fig.3.

Merging branch histories is not performed if all of their first terms are *unknown*. If these branch histories include a branch history in which the first term is fixed (known), the possibility of merging would be checked. For example, if there exist two $< unknown, 2, unknown >$s in T, merging these branch histories into $< 2, unknown >$ will occur when $< 3, 2, unknown >$ is received. If there exist $<$

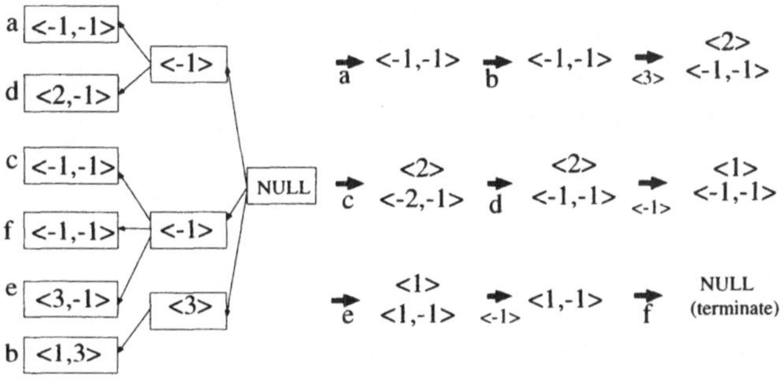

Figure 3: Packet level retrieval termination detection

$unknown, 2, unknown >$ and $< 3, 2, unknown >$ in T, when $< unknown, 2, unknown >$ is received, these three branch histories will also be merged into $< 2, unknown >$. The changes to T are shown in the right-hand side of Fig.3. Note that merging is impossible even if c is received and is postponed until d having the fixed first term of the branch history will be received.

Lastly, we estimate the overhead of communication cost due to attaching branch histories. Let n be the number of the objects retrieved in a PE, O_S be the size of an OID, H_S be the size of a branch history, and P_S be the size of a packet. Whereas object-level termination detection requires a branch history for each object, packet-level detection requires a branch history for each packet. Let β, β_o, and β_p be the number of packets required in no termination detection, object-level detection, and packet-level detection, respectively. For example, let O_S be 8 bytes, P_S be 256 bytes, and H_S be 12 bytes. The required numbers of packets can be calculated as shown in Table 1, which indicates that the communication overhead caused by the packet level detection is very small.

Table 1: Number of required packets

n	10	20	40	100	300	500	1000
β	1	1	2	4	10	16	32
β_o	1	2	4	9	25	42	84
β_p	1	1	2	4	10	17	34

3 A database programming language based on the object-parallel approach

Parallelism can be exploited not only in the retrieval phase of the database management system (DBMS), but also in the application processing phase after the retrieval. If we can develop a parallel object-oriented programming language for database applications, we will be able to easily write portable programs for non traditional database applications such as a direct simulation over a large number of CAD DB objects.

In such database applications, the number and topology of objects to be processed by an application program cannot be determined prior to run time. However, most parallel programming languages designed and developed for scientific computation are based on the *data-parallel approach*[9)] and therefore assume the number and topology of objects are fixed at compilation time.

In order to overcome this problem, an *object-parallel approach* for database programming languages is proposed. Unlike most data-parallel programming languages, the proposed language can execute parallel operations over an arbitrary set of objects and supports object references through object identifiers (OIDs). Since there is no restriction on the number or topology of the objects to be processed in parallel, application programs can receive objects that satisfy a certain condition and have complex reference structures among these objects.

Basic Features

In the proposed language, an exclusively parallel construct, designated a `for all` is designed for parallel operations. The basic syntax is defined as follows.

`for all` *class variable* `in` *union*
`[such that` *condition* `]`
`do` *method-invocation*

The above lines apply a *method-invocation* (or a sequence of multiple method invocations) over an arbitrary *union* of sets of objects pointed by *variable* in parallel, where in a *such that* clause can be used as a filter to eliminate irrelevant objects in the union.

Programmers need not consider the number or topology of objects to be processed. Rather, the union of object sets and the operation to be applied to those objects need only be specified. The object sets are actually distributed over processing elements (PEs) of a distributed-memory parallel computer such as a workstation cluster.

However, the number and topology of objects cannot be determined at compilation time in database applications. Therefore, this language requires the following two features.

1. A location-independent object reference mechanism
   Since database objects are not packed in an array and the relationships among objects are expressed by OIDs, the reference mechanism cannot be based on array subscripts. In addition, the reference mechanism must also be location-independent, because the next feature requires object relocation at run time.

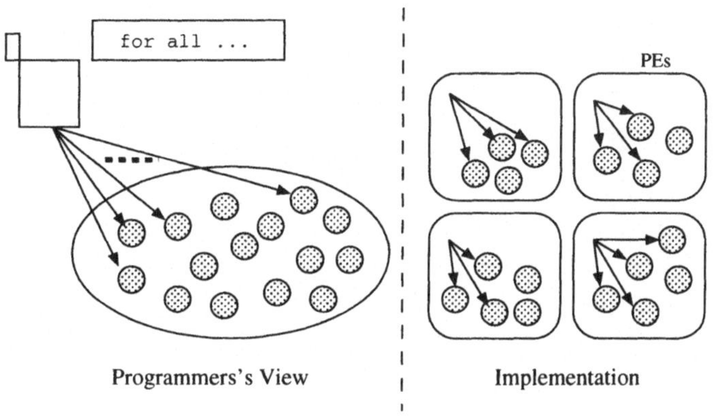

Figure 4: Diagram of `for all` statement

2. A dynamic performance optimization mechanism
   Unlike array computation, the load information cannot be estimated at compilation time. The information must be collected at run time, and the necessary actions must be taken when the load imbalance is excessive.

Location-Independent Object Reference Mechanism

The location-independent object reference mechanism is implemented within a *heap management object* (HMO) allocated in each PE[10)]. An HMO can translate an abstract OID to its address when the object referred by the OID is local, or can obtain the ID of the PE holding the object when the object is remote.

An HMO manages all incoming messages and outgoing messages. The HMO looks up its *object reference table* (ORT) whenever an OID reference is found in an operation, and determines whether the OID is located within the PE. If the target object is not registered in the ORT, the object is remote and the HMO sends an inter-processor communication to delegate the operation to another PE.

When the HMO cannot find the entry for the given OID, the HMO looks up its *object allocation table* (OAT) and checks whether the object is registered. If the object is registered, the HMO sends a 1-to-1 message to the PE. If not, the HMO broadcasts a request message and registers the PE ID after receiving the reply from the remote PE holding the object.

This indirect reference mechanism is sufficiently effective even for complex operations such as traverses[10)].

Since the individual OIDs contain no location information, objects can be relocated without destroying the relationships among them. This feature is essential for dynamic performance optimization mechanisms.

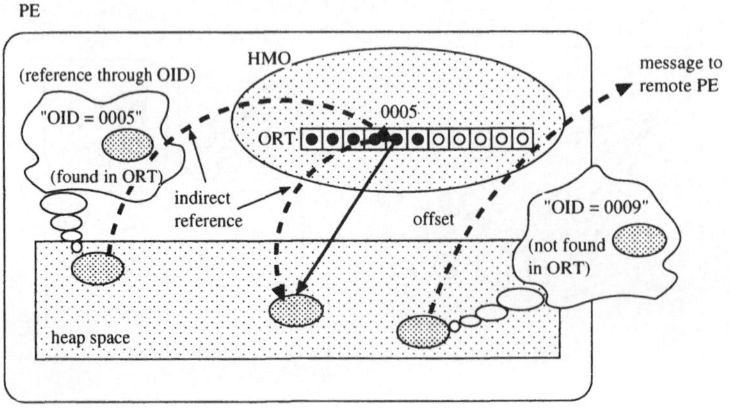

Figure 5: Object Reference Table and References to Local Objects

Dynamic Performance Optimization Mechanism

The dynamic performance optimization mechanism has been designed as a suite of dynamic object relocation mechanisms. These relocation strategies are based on approximate load information, which can be collected without significant overhead at run time.

The basic strategies are categorized as follows[13)].

1. Object count balancing

   This strategy relocates objects so that the number of objects on each PE becomes approximately equal. Object count balancing is quite simple since this strategy only compares the maximum and minimum numbers of objects allocated to PEs and relocates surplus objects to the PEs that have fewer objects.

2. External reference reduction

   If a certain object has a larger external reference frequency, that object is relocated to the PE that accesses it most frequently.

   For each object in each PE, this strategy requires additional storage space in order to collect the information necessary for determining the relocation candidates and the relocation destinations.

3. Locality preference enforcement

   If a certain type of reference among objects is known to be accessed frequently, a pair of objects linked by that special reference should be allocated on the same PE. This strategy corresponds to object clustering.

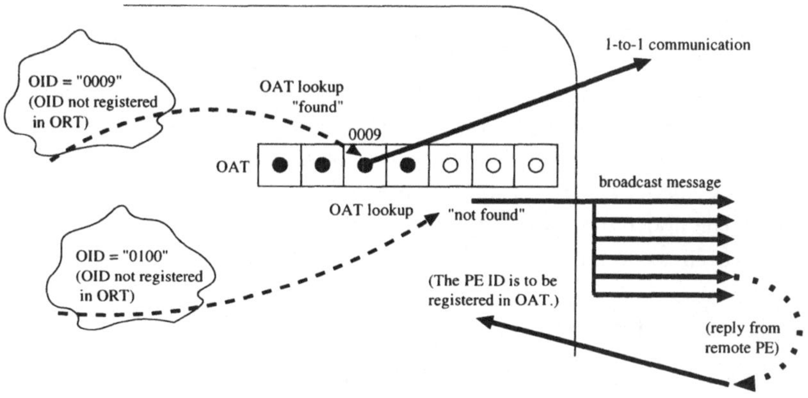

Figure 6: Object Allocation Table and References to Remote Objects

Object count balancing and locality preference enforcement can be implemented without significant space/time overhead[12)].

For external reference reduction, however, a simple implementation based on "most-recently-accessed PE" is not effective[12)]. Therefore, another scheme has been proposed which identifies the relocation candidates and the destinations more precisely[8, 11)].

In order to find the relocation candidate objects more precisely, four counters are introduced for each object. If the relocation timing arrives, the following formula selects the candidates for relocation:

$$R_{to\_inner} + R_{from\_inner} < R_{to\_ext} + R_{from\_ext}$$

where R_{to_inner}, R_{from_inner}, R_{to_ext} and R_{from_ext} are the references to the same PE, the references from the same PE, the references to the other PEs, and the references from the other PE, respectively.

For the problem of finding the destination PE, n records (n is a constant, possibly smaller than the number of PEs) is introduced to count the external references more precisely. Each record has PE ID and the reference counts from the PE. Let the i-th record be record(i) ($1 \leq i \leq n$). The revised method selects the PE having the largest count as the relocation destination.

The counts must be initialized to zero. The revised algorithm for updating the counters, to be initiated each time a reference is encountered, is as follows.

1. Look up the PE ID recorded in record(1) and increment the counter if the PE ID matches the ID of the referrer. If not, go to Step 2.

2. For each record(i) ($2 \leq i \leq n$), repeat the following operation. If the PE ID matches the ID of the referrer, increment the counter. If the counter for record (i) becomes greater than that of record($i-1$), replace the record with record($i-1$). Repeat this step until the matching record is found or until $i = n$.

3. If the matching record is not found for $1 \leq i \leq n$, overwrite the record(n) with the referrer ID and its reference count 1.

The above data structure can be duplicated to achieve better performance. The new structure records the PE ID and the references *to* the PE in a manner that is symmetric to the previous structure. By using this information, the object relocation can successfully decrease the external references.

Although the above-described scheme is still an approximation, the performance has been reported to be improved[8, 11].

Future Research Issues

The object-parallel approach for parallel database applications has several advantages over conventional data-parallel approaches. However, a number of questions remain that must be answered in future studies.

First, the interactions among the object relocation strategies should be investigated. When studying the object relocation strategies, each strategy was tested separately. If applied at the same time, however, some objects may experience conflicting relocation operations caused by multiple strategies, and may be relocated in an improper manner.

Second, a new generation of high-performance parallel machines should be considered. Due to recent advances in hardware technology, a symmetric multiprocessor (SMP) has become a reasonable choice for a PE. An SMP cluster, SMPs connected via a high-speed network, is already a feasible solution for a high-performance database server. In this configuration, the object reference mechanism and the object relocation mechanisms may require revisions.

4 An object database management system on a workstation cluster

ShusseUo is a scalable distributed and parallel object database management system[17] for a *workstation cluster.* The object database system is effective for advanced applications such as multimedia databases, and engineering databases, because complex-structured data can be represented naturally using an object model, such as *OID* or *relationship*, which are given by an object database. Client programs share a database and often send a request to a server concurrently. A workstation cluster is a set of distributed computers connected by computer network(s). Database server programs run on all sites of a workstation cluster, and the workstation cluster is expected to act as one parallel database machine. An ideal parallel machine has linear scalability as the number of sites increases.

Architecture

ShusseUo is an ODMG3.0-compliant object-oriented database system. The Object Database Management Group (ODMG) produced a standard for object database management systems which allows application developers to write portable applications[14]. The standard when we started development of ShusseUo was ODMG .0, and the most recent version is release 3.0, an enhancement of ODMG 2.0.

The ODMG architecture defines a data definition language, a query language, and a number of manipulation languages. The names of the languages are Object Definition Language (ODL), Object Query Language (OQL), and Object Manipulation Language (OML), respectively, and these are based on a common object model. OML has three language bindings: C++, smalltalk, and Java.

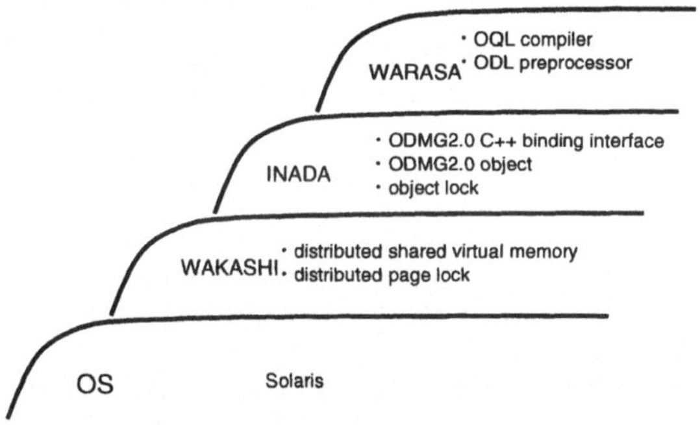

Figure 7: Architecture of ShusseUo

ShusseUo consists of server, client library, and ODL/OQL programs, which are named WAKASHI, INADA, and WARASA, respectively (see Figure 7).

WARASA consists of the ODL preprocessor and OQL compiler programs. The ODL preprocessor is the language for schema definition. The OQL compiler reads embedded OQL statements in a C++ binding OML program and compiles them to equivalent C++ binding OML statements. Figure 8 shows an example of an embedded OQL statement.

```
d_Ref<d_Bag<d_Ref<Person> > > result;
d_OQL_Query q ("select p from Persons  p \
    where p .name =\"Smith\"");
d_oql_execute(q,result);
```

Figure 8: C++ OML program code containing one embedded OQL statement

INADA is for use with C++ OML. INADA is a class library that provides an ODMG 2.0 C++ binding OML, and ODMG C++ objects. INADA is included in each client program. WARASA programs also include INADA.

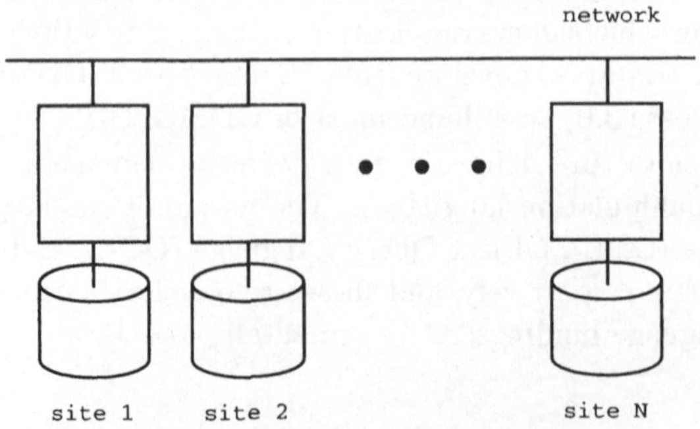

Figure 9: Workstation cluster

WAKASHI is a server program that runs on each site of a workstation cluster, which is a set of computers connected by a computer network (see Figure 9).

ShusseUo is a multi-client multi-server system. WAKASHI provides the services of the database server to INADA. The main service of WAKASHI is a data storage space named *heap*. WAKASHI manages heap, and has functions of distributed locking, distributed caching, transaction commit and abort, logging, and recovery. A client program directly maps a heap provided by WAKASHI onto a part of the virtual memory space of the client program. Then, the client program reads and writes data in heap by memory address.

WAKASHI is a page server and does not depend on the data model or query language. Heap is a global data storage space in a workstation cluster. All of the client programs on a workstation cluster share the data in the heap. Heap has fixed-length and an array of pages. WAKASHI receives a page number and lock mode from client programs for distributed locking and uses this page number for data transfer between distributed servers. The operations of heap are open, close, map, lock, commit, and abort. The lock operation is used to obtain a read page lock or a write page lock. WAKASHI transfers a data page when the page lock is obtained and writes the data page back to a database file upon transaction commit.

Data stored in a heap is persistent when the heap is persistent. A heap is persistent when the heap is mapped onto a database file. The mapping is disk-memory mapping. WAKASHI uses the memory-mapped file OS service for disk-memory mapping. WAKASHI maintains the before images of the pages written by client programs and uses these for transaction abort. In summary, heap is an extension to distributed shared virtual memory, and a heap is persistent when the heap is mapped onto a database file (see Figure 10).

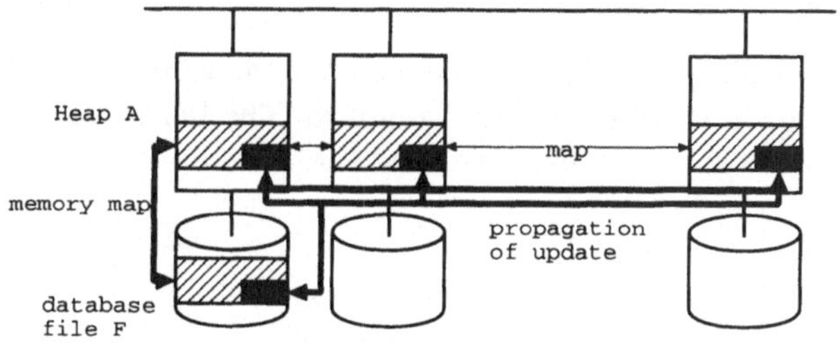

Figure 10: A persistent distributed shared virtual memory

Caching and clustering

WAKASHI is for distributed and parallel processing of a database. The database files should be distributed over distributed sites in order to increase the total system performance. The storage manager of ShusseUo, WAKASHI controls the concurrent accesses to database files and manages the distributed cache of heap. Once a data page of a heap is transferred, the page data is cached until the page data becomes invalid by as the result of a page write.

INADA is a class library that provides an ODMG 2.0 C++ binding OML and ODMG C++ objects and provides services by which a client program manages persistent objects. The interface for object creation is an object constructor with a *new* operator, and the *delete* operator is provided for object deletion. INADA receives a request to create or destroy an object and allocates or releases a memory block in a persistent heap.

Persistent heaps can be distributed, and so persistent objects can be distributed. The two types of object distribution are as follows.

- dynamic distribution:

  Persistent objects are stored in heaps in one site. WAKASHI allows programmers to write location transparent client programs. The same client program can run on a different site because copies of the data pages that contain object data are transferred and cached automatically by the WAKASHI mechanism at program execution.

- static distribution:

  Persistent objects are stored into distributed heaps. Objects are distributed at their creation. The network cost can be reduced if the distribution of objects and the execution site of the program are carefully designed.

Preliminary Benchmark

We use an extended SEQUIOA 2000[19)] benchmark to evaluate ShusseUo. The data used in the benchmark is geographically complex. The benchmark was designed to evaluate an object database system. We used the *dynamic distribution* approach in our benchmark.

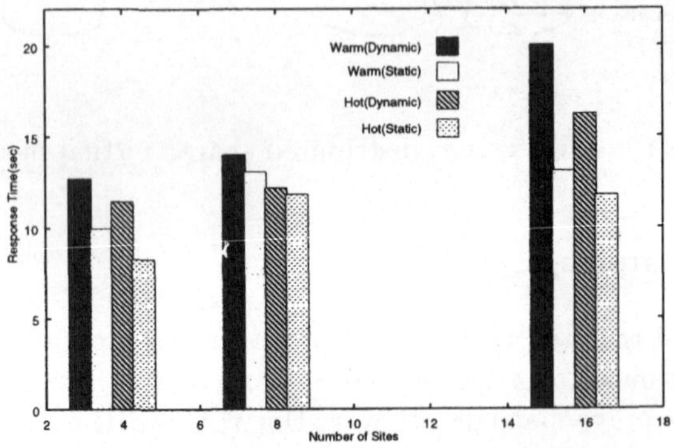

Figure 11: Scale-up test results

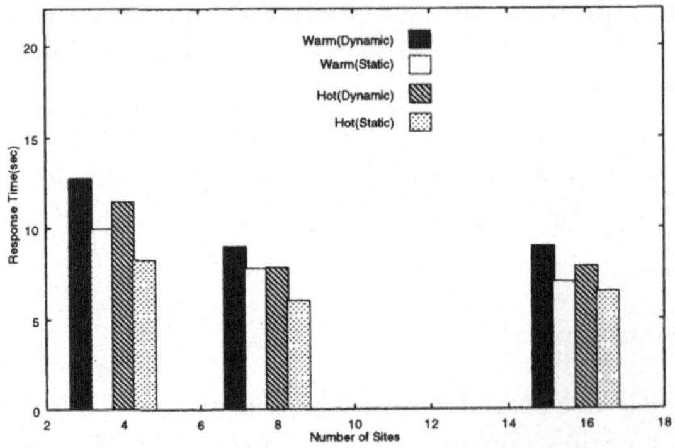

Figure 12: Speed-up test results

Figures 11 and 12 show the results of speed-up and scale-up tests, respectively, as the number of sites is changed from 4 to 8 and 16. The same operation was repeated ten times using different query conditions. The query condition is indicated by the position of the rectangles. The average response time among the ten tests is shown as warm result in Figures 11 and 12. After the warn test, we performed the same series of tests to obtain the hot result. The hot result is faster than the warm result in this test because in the hot tests, the copies of the page data are cached at each execution site by WAKASHI. The test results shows that the dynamic distribution is good because the operation is read-only and the page data is cached at each site.

Conclusion

ShusseUo is the first trial of an object database system on a workstation cluster. The server of ShusseUo, WAKASHI, is a process that runs on each site of a workstation cluster and communicates with other sites. WAKASHI manages *heap*. The heap area in the server virtual memory space is mapped onto a database file (*disk-memory mapping*), onto other heap areas on other servers, and onto client virtual memory spaces (*memory-memory mapping*). This mechanism allows the writing of location transparent client programs as well as dynamic distribution of databases. The hot result achieved by our benchmark indicates the effectiveness of dynamic distribution and caching.

Acknowledgments

The present study was supported in part by the Grant-in-Aid for Scientific Research (10308012) from the Ministry of Education, Culture, Sports, Science and Technology of Japan.

Bibliography

1) M. Nagata, "Parallel Processing for Indexed Structure of Objects", Proc. of the Int'l. Symposium on Next Generation Database Systems and Their Applications, pp. 56–61,1993.

2) E. Bertino ,"A Survey of Indexing Techniques for Object-Oriented Database Systems",in Query processing for Advanced Database, Freytag, J.C., Maier,D and Vossen, G.(Eds.), pp.193–209, 1995.

3) A.K. Thakore, S.Y.W. Su, "Algorithms for Asynchronous Parallel Processing of Object-Oriented Databases", IEEE Trans. on Knowledge and Data Engineering, Vol. 7, No.3, pp.487–504, 1995.

4) G.S. Chinchwadkar, A. Goh, "Transforming Complex Methods in Vertically Partitioned OO Databases", Proc. of Int'l Conf. of Parallel and Distributed Computing and Networks, pp.56–59, 1997.

5) L. Mutenda, M. Hiyama, T. Yoshinaga, T. Baba, "Parallel Navigation in an A-NETL Based Parallel OODBMS", Proc. of Int'l Symposium on High Performance Computing, pp.305–316, 1997.

6) K. Ogura, T. Tsuji., A. Vreto., T. Hochin, "A Method of Horizontal and Vertical Index Splitting for Complex Objects", Transaction of IEICE, Vol.J80-D-I,6,pp. 486–494,1997.

7) T. Tsuji, K. Higuchi, K. Ogura, T. Hochin, "Detection of Parallel Retrieval Completion for Complex Object Index", Transaction of IEICE, Vol.J82-D-I,No.2, pp.446–449, 1999.

8) H. Amano, K. Kimura, and A. Makinouchi, "Improving Dynamic Load Balancing Mechanisms for a Massively Parallel Database Programming Language," In *Proc. Database Applications in Non-Traditional Environments '99,* The IEEE Computer Society, 2000.

9) P. J. Hatcher and M. J. Quinn, *Data-Parallel Programming on MIMD Computers,* The MIT Press, 1991.

10) K. Imasaki, T. Ono, T. Horibuchi, A. Makinouchi, and H. Amano, Design and Evaluation of the Mechanism for Object References in a Parallel Object-Oriented Database System," Proc. Int. Database Engineering and Applications Symp. (IDEAS'97), Montreal, Quebec, pp.337-346, Aug. 1997.

11) K. Kimura, H. Amano, and A. Makinouchi, Dynamic Performance Optimization for Parallel Object-Oriented Database Programming Languages, Proc. the 2000 International Database Engineering and Applications Symposium(IDEAS), Yokohama, Japan. Sept. 2000 (to appear).

12) T. Ono, H. Amano, T. Horibuchi, and A. Makinouchi, Performance Evaluation of Object Relocation Mechanisms for Dynamic Load Balancing in Parallel Object-Oriented Databases, Proc. Int. Symp. on Information Systems and Technologies for Network Society, Fukuoka, Japan, pp.352–355, Sept. 1997.

13) T. Ono, T. Horibuchi, K. Imasaki, H. Amano, and A. Makinouchi, Design of Dynamic Load Balancing Mechanisms for Massively Parallel Object-Oriented Databases, In *Cooperative Databases and Applications (Y. Kambayashi and K. Yokota eds.),* pp.569–572, World Scientific, Singapore, 1997.

14) Botao Wang, Hiroyuki Horinokuchi, Kunihiko Kaneko, Akifumi Makinouchi, Parallel R-tree Search Algorithm on DSVM, Proceedings sixth International Conference on Database Systems for Advanced Applications (DASFAA 99), pp.237-246

15) R.G.G. Cattel et.al. "Object Database Standard ODMG2.0," Morgan Kaufmann Publishers, 1997

16) Richard Cooper, "Object Databases: An ODMG Approach," Morgan Kaufmann Publishers, 1997

17) G.Yu, K.Kaneko, G.Bai, A.Makinouchi, "Transaction Management for a Distributed Object Storage System WAKASHI – Design, Implementation and Performance," 12th Int'l Conf. on Data Engineering, 1996.

18) Qiang Fang, Guoren Wang, Ge Yu, Kunihiko Kaneko, and Akifumi Makinouchi,

Design and Performance Evaluation of Parallel Algorithms for Path Expressions in Object Database Systems on NOW, 1999 International Symposium on Database Applications in Non-Traditional Environments(DANTE'99), pp. 373 – 380

19) M. Stonebraker, J.Frew, K.Gardels and J.Meredith, The SEQUOIA 2000 storage benchmark, SIGMOD 1993.

Index